The Lincoln Highway

The Lincoln Highway

Main Street
across America

Text and
photographs by
Drake Hokanson

Ψ

University of Iowa Press
Iowa City

University of Iowa Press, Iowa City 52242

Printed in the United States of America

Second printing, 1988

Book and jacket design by Richard Hendel

Typesetting by G&S Typesetters, Inc., Austin, Texas

Printing and binding by BookCrafters, Chelsea, Michigan

Library of Congress Cataloging-in-Publication Data

Hokanson, Drake, 1951–

The Lincoln Highway: main street across America /

text and photographs by Drake Hokanson.—1st ed.

p.   cm.

Bibliography: p.

Includes index.

ISBN 0-87745-197-4

1. Lincoln Highway—History.   2. Lincoln Highway—

Pictorial works.   I. Title.

HE356.L7H65 1988   87-30167

388.1′0973—dc19   CIP

For Carol,

who knows every mile, every word,

every photograph, and who em-

bodies the grand spirit of "the open

road and the flying wheel."

Publication of this book

was supported in part by a

generous grant from the

University of Iowa Foundation

Contents

Acknowledgments

This book, like any other, bears the invisible marks of many individuals. By far the most helpful in a material sense (and many times in a spiritual sense as well) were the many librarians, assistants, and researchers who bent frontwards and backwards to find information of all kinds. Many who helped never even whispered their names, but others are herein named: W. David Lee of the American Association of State Highway and Transportation Officials; Sue Williams of the American Automobile Association Library; Jane Pollock and Sharon Balius of the University of Michigan Engineering and Transportation Library; Hank Zaletel of the Iowa Department of Transportation; O. H. Van Zee of Arlington, Virginia; and Richard Menzies, Salt Lake City. Special thanks goes to Richard F. Weingroff of the Federal Highway Administration for his eleventh-hour photocopying spree, and to friends and librarians Margaret Richardson and Robert McCown here at the University of Iowa.

There are many, many people who have an abiding interest in the Lincoln Highway, in old highways generally, and in travel in America. Some have recently driven part of the highway for the first time; others have made coast-to-coast trips a lifelong pursuit. Some are local historians or residents of communities along the route; others are antique automobile collectors, Oregon Trail buffs, or people who found something in the attic that they thought might be of value to me. Many people sent me diaries, guidebooks, postcards, maps, photographs, clippings, and splendid tidbits of information. Others along the route fed and housed me, helped me find old routes, or stood patiently while I took their photographs. All freely offered their time, information, and enthusiasm.

To the hundreds who wrote or called whom I haven't space to name, my grateful thanks and apologies; to others, a special thank you:

David and Reuvo Bagley, James H. Baxter, Duane Best, Joseph A. Carpentier, Mrs. Joseph DeHart, Rick Donaldson, George C. Eppinger, John Fondersmith, A.W. Gaines, Josephine E. Hayden,

Howard G. Henry, Frances M. Herndon, Max E. Hobbs, Pam Kocha-nowski, Pauline Lea, David B. Lockard, Dick Netzer, Richard L. Pad-dock, Roy Sankot, Raymond, Anna, and Dale Sipe, Bradley Skinner, R. F. S. Starr, and last, but never least, Marvin Wolfe.

Others contributed in material ways. Lyell Henry lent a deep pile of Lincoln Highway postcards; editor Greg Gullickson lent grammar and style to a rough manuscript. Charles Roberts cleared twenty pages in *The Iowan* for my story about the Lincoln that got this whole thing moving, and Drs. Alan Hathaway and J. K. Johnson provided maps, names, and ideas.

And last, three individuals deserve special recognition for both ma-terial and spiritual assistance: friend and unpaid editor Charlie Drum; friend and enthusiast David Plowden; friend, reader, and lifetime trav-eling companion Carol Kratz. Without Charlie, the manuscript never would have taken shape; without David, I might have followed a dead-end road; without Carol, I never would have started.

The Lincoln Highway

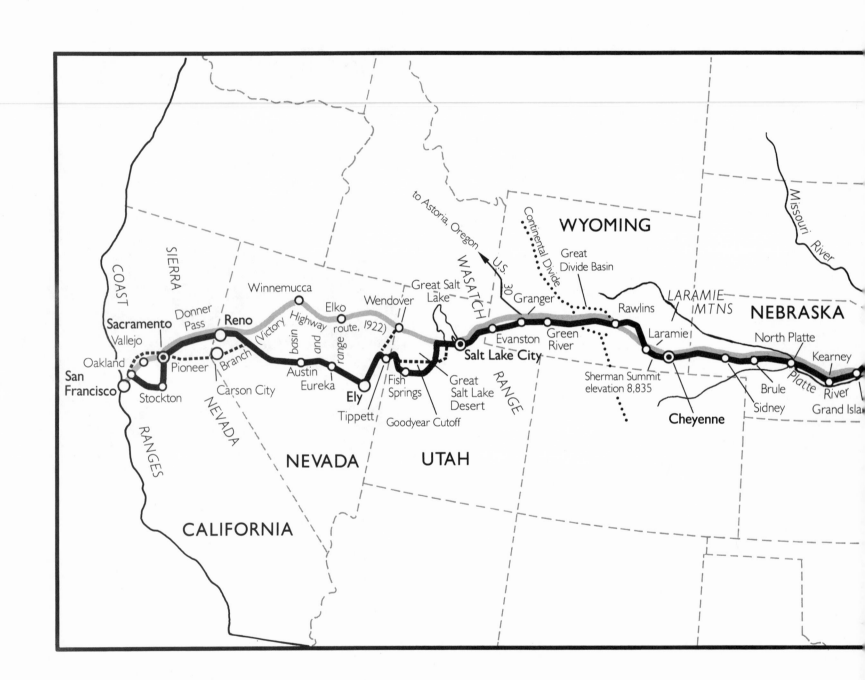

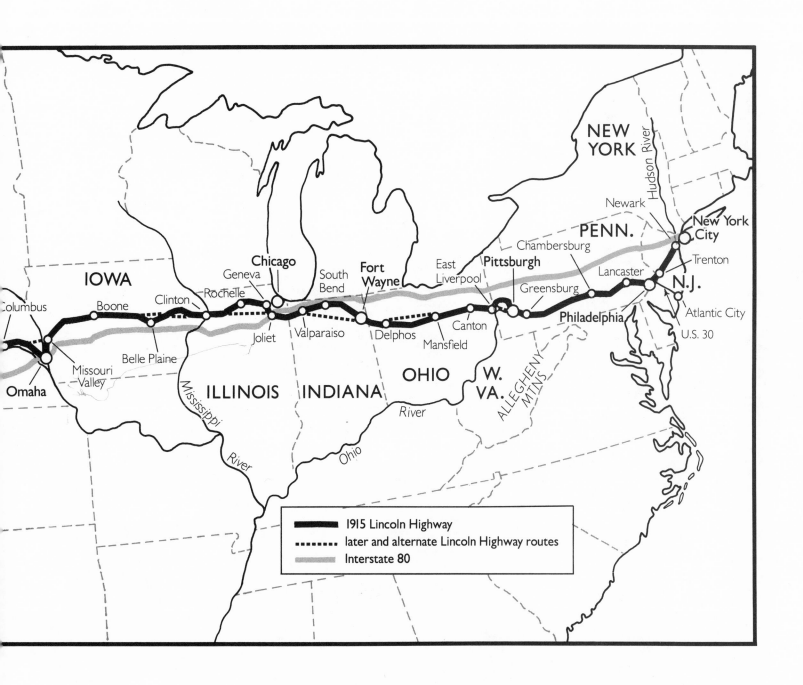

IOWA
Columbus
Boone
Clinton
Rochelle
Geneva
Chicago
South Bend
Fort Wayne
East Liverpool
Pittsburgh
Chambersburg
NEW YORK
PENN.
Newark
New York City
Trenton
Lancaster
N.J.
Greensburg
Philadelphia
Atlantic City
U.S. 30
Belle Plaine
Joliet
Valparaiso
Delphos
Mansfield
Canton
W. VA.
ALLEGHENY MTNS
Missouri Valley
Omaha
ILLINOIS
INDIANA
OHIO
Mississippi River
Ohio River
River
Hudson River

	1915 Lincoln Highway
	later and alternate Lincoln Highway routes
	Interstate 80

Introduction

In the early days of automobile travel, before transcontinental journeys were commonplace, it was a ritual of American adventure to dip your tires in one ocean before starting out for the other. Then, even though you probably changed tires several times en route, you could dip them in the farther ocean when you arrived there, certifying beyond question that you had traveled the entire distance, that you had in some way connected the Atlantic with the Pacific. If you were afraid of getting your car stuck on the beach, a variation was to fill a bottle with the brine of one ocean to carry across the continent to dump in the other. In either event, the arrival at the far ocean was cause for celebration, cause for a certain kind of pride in personal accomplishment and in a nation that extended from sea to shining sea.

For almost two hundred years, Americans have been fascinated by this notion of tying one coast to the other. Whether through national land acquisition, the building of transcontinental institutions that made travel or communication easier, or through individual journeys, we have been fixed on the idea of spanning the continent. The excitement started in the dawning years of the nineteenth century with the Louisiana Purchase and the Lewis and Clark expedition, and was a part of national pride by the time California was admitted to statehood in 1850. By then the nation's land holdings had filled in most of what makes up the lower forty-eight states today. The nation had tripled in size in fifty years, and Americans who had looked westward from Kentucky found themselves standing on the far coast and looking westward out over the Pacific.

What early auto travelers felt as they dipped wheels in oceans and crossed the continent must have been a vestige of Manifest Destiny, that great rationalization that allowed the nation to claim, buy, and steal all the miles of land to the Pacific and beyond. The notion was so ingrained, so much a part of the national consciousness by the turn of the century, that any traveler seeing the blue line of one ocean after starting from the other must have felt a personal sense of national fulfillment.

The first continental crossing by automobile came in 1903. That germinal year also saw the incorporation of the Ford Motor Company and the first flight by the Wright brothers. About the middle of May, a Vermont physician accepted a challenge to drive an automobile from San Francisco to New York. Dr. H. Nelson Jackson never got rich from the bet; the stakes were but fifty dollars. He hired Sewell K. Croker as his mechanic, and on his advice bought and equipped a used twenty-horsepower, two-seat, open-top Winton. On May 23, four days after making the bet, they left for the East. Near Caldwell, Idaho, the pair adopted a stray bulldog whom they named Bud. As they had done for themselves, Jackson and Croker provided their new companion with driving goggles to protect his eyes from flying mud and blinding dust.

The trio traveled some six thousand hard miles and arrived in New York City in the wee hours of July 26, sixty-five days after leaving San Francisco. In all, they spent some three weeks of the time repairing the car, resting, or waiting for parts. For about half the distance they traveled what would in ten years hence become America's first transcontinental highway.

The Lincoln Highway was an expression of the national desire to bind the country from east to west. It captured the American imagination in much the same way Jackson's feat did, and in the same fashion as the great westward migration on the Oregon and California trails, the pony express, and the transcontinental railroad had a half century earlier. Along with the motorcar, the Lincoln Highway allowed ordinary citizens the opportunity to follow Jackson's trail, to make their own journey, to express their own transcontinental aspirations, to carry their own bottle of water from one shore to the other.

This, then, is a story about an American highway that helped teach us a new way to travel. It is about the men who tried to see it built; it is about how this highway faded when newer ideas came along. It is a story about the geography of travel across 3,300 miles of the United States, and it is a tale of some who made the trip.

I

The Early Highway

1

An Imaginary Line
Like the Equator!

Emily Post sat high in the rear seat of the open touring car as they drove up Fifth Avenue. She wore a large hat to keep the warm spring sunlight out of her eyes and held a large bunch of violets, a farewell gift from her editor. The car was heaped with dunnage sufficient for a world cruise. Her son Ned was at the wheel, and a cousin, Alice Beadleston, sat buried under luggage.

Near Forty-second Street, a few blocks from Times Square, they met many friends, also in cars, who were curious about this entourage so obviously set for a long trip. Respected members of New York society, the Posts wanted to be seen as they left the city, and so took every opportunity to announce that they were not headed for the Berkshires or the Adirondacks or the resorts of Maine; they were headed clear across the continent, across some three or four thousand miles of Parts Unknown, headed for San Francisco and the Panama-Pacific International Exposition. The year was 1915.

The trio left New York City several hours late. They had planned to leave by nine that April morning, but they did not finish packing and loading until early afternoon. As with nearly every automobile trip before or since, departure was delayed because it took much longer than expected to decide about last items and to finally load and arrange the car.

The choices about what to take and what to leave were of great concern. Emily was an experienced traveler, but her automobile travel experience was limited to tame adventures along the paved roads of Europe, where class hotels and fine restaurants awaited around each bend. In the States, she traveled by luxury train or steamship. She and Ned had toured Europe by auto the year before, just as the war was beginning, "but our own land," she wrote, "except for the few chapter headings that might be read from the windows of a Pullman train, was an unopened book."

What should one take on a journey across America, across endless mountain and desert, across broad expanses where small-town commercial hotels, tearooms, and poor cafes would be the only fare? Like most Americans, Emily Post was a complete stranger to the American road. She simply had no experience to prepare her for a journey across the continent by automobile.

With characteristic zeal, she had earlier sought all available information about routes, conditions, and accommodations. What she had found was not encouraging. A young man in the touring department of a New York auto club tried to convince her that a tour of the Berkshires was really a much better idea. Maps and guides for New England were plentiful, but for the West, well, those were scarce and hardly accurate. Finally she found a map showing four equally straight and inviting routes to the Pacific Coast. A well-traveled friend was consulted. Post wrote:

"Can you tell me," I asked her, "which is the best road to California?"
Without hesitating she answered: "The Union Pacific."
"No, I mean motor road."
Compared with her expression the worst skeptics I had encountered were enthusiasts. "Motor road to California!" She looked at me pityingly. "There isn't any."
"Nonsense! There are four beautiful ones and if you read the accounts of those who have crossed them you will find it impossible to make a choice of the beauties and comforts of each."
She looked steadily into my face as though to force calmness to my poor deluded mind. "You!" she said. "A woman like you to undertake such a trip! Why, you couldn't live through it! You don't know what you are undertaking."
"It can't be difficult; the Lincoln Highway goes straight across."
"In an imaginary line like the equator!" She pointed at the map that was opened on the counter. "Once you get beyond the Mississippi the roads are trails of mud and sand. This district along here by the Platte River is wild and dangerous; full of the most terrible people, outlaws and bad men who would think nothing of killing you if they were drunk and felt like it. There isn't any hotel. Tell me, where do you think you are going to stop? These are not towns; they are only names on a map, or at best two shacks and a saloon! This place North Platte why, you couldn't stay in a place like that!"

Discouraged, Post consulted another touring agency. Here the picture was a little more promising. Yes, there were roads, yes, there were hotels, but few were what could be called luxurious. No, there wasn't much pavement, or even much gravel, especially west of Chicago—just good natural dirt road—but you could make good time, if it didn't rain. And if it rained? You could get out your solitaire pack and wait for sunshine. Post was also told that the Lincoln Highway across the mountains of Pennsylvania was under construction and unfit for travel. She was advised to go via Albany and Cleveland and to pick up the Lincoln Highway at Chicago. Post was starting to have doubts about this much-touted boulevard, and she was disappointed that they wouldn't be driving it right from its beginning in New York City.

Her gloom, however, turned quickly to excitement as they motored north along the Hudson River. The Lincoln Highway could wait a few miles.

Emily Post watched the passing scenery and reveled in the freedom and exhilaration felt by auto travelers of the day. The cramped city was far behind; she now had the wind at her face and the open road ahead. She also had escaped the crowded trains and their limiting schedules. They could stop when and where they wanted, need talk only with whom they pleased. They rolled swiftly along the Hudson River in the open car. Ahead lay thousands of miles of yet unseen road, new acquaintances, unimagined experiences. The day was warm; their hand-built European car rumbled strongly through rakish side pipes. This was no flivver! The scents of river, woodland, and field came and went as they motored along. Post was disappointed that no one cheered as they passed, or wished them well on their journey to the Pacific.

When teatime arrived that first afternoon, the travelers were well into the countryside, away from the city and noise, and were beginning to feel the pull of the road. But teatime was teatime, and Ned was instructed to pull off the road. Emily wanted to make tea and to try out an elaborate silver tea service given by a friend as a farewell present. It had come the day before in an enormous wicker picnic basket and apparently was intended for serving high tea among the cowboys. The basket had arrived packed with bottles, plates, jars, boxes, flat silver, cutlery, the tea set, and a note saying, "I could not bear to think of you starving in the desert." Crossing the continent by automobile was one thing, but giving up the true comforts of home was quite another. It took several minutes of digging and unloading to uncover the basket, and several more to locate sandwiches and tea-cakes within. Then, along the roadside, the women carefully brewed tea while Ned looked after the car and thought of the road ahead.

But to understand what would happen to Emily Post, to understand what she and others would find along the Lincoln Highway, we must look at a time two and a half years earlier: September 1912. Up until that time, very few people had ever seriously considered driving an automobile across the country. But during that month Carl Graham Fisher hatched a new idea that would help change how America traveled.

Fisher was a man of ideas, and big ideas at that. His Indianapolis Motor Speedway had become a smashing success, especially after he paved it with brick and inaugurated the Indianapolis 500 in 1911. He was founder of the Prest-O-Lite Company, maker of carbide headlights, which in the days before the all-electric car provided the only reliable light for night driving. He was an athlete, a keen lover of

Carl G. Fisher.
Courtesy of the University of Michigan.

yachting and ballooning, and never one to pass up a good promotional idea. In 1908, as a publicity stunt, he attached a Stoddard-Dayton auto to a helium balloon and floated across Indianapolis. He had been a bicycle salesman and racer, and later turned to racing automobiles, at which he broke a few world records and raced against the famed cigar-chomping Barney Oldfield. Later in life he would be the founder and promoter of one of the biggest land booms in the country: Miami Beach, Florida.

But in the early fall of 1912, Fisher's old ideas, now successful on their own, must have seemed a little stale, and he had a new scheme. He was restless, like a man with an urge to wander, but instead of dreaming about making a trip himself, he dreamed of making a road so that others could.

His new idea was called the Coast-to-Coast Rock Highway. Realizing that the success of any idea depended a great deal on rousing the enthusiasm of others, he disclosed the plan to anyone who would listen. In early September of 1912, he threw a large dinner party for the leaders of Indianapolis automobile manufacturing. Fisher knew that the enthusiasm and capital of these men could get this highway started. After dinner at the Old Deutsches Haus, he spoke of the need for good roads and laid out his plan. "A road across the United States!" Fisher shouted to the group. "Let's build it before we're too old to enjoy it!"

He told them that a graveled highway could be built from coast to coast for about ten million dollars, a low figure even for 1912. This money would be used to buy only basic road-building materials; the labor and machinery, he said, would be provided by the counties, towns, and cities along the yet undetermined route. What community could turn down the opportunity of free materials and a place on the map astride America's first transcontinental highway? In addition, Fisher wanted this dream highway finished in time for twenty-five thousand cars to cross the continent for the Panama-Pacific Exposition only two and one half years away.

By the time of his highway idea, Fisher had made his millions, though he would later lose them in Florida. He was intent on seeing this highway built not to earn him money, nor to gain attention for himself, but because he felt the nation needed it.

Because of the exposition, San Francisco would be the western end point for the highway, and there was little doubt that America's great metropolis, New York City, would be the eastern terminal. Fisher would not divulge any details about the route between these points; in fact, no route had been selected or even considered. He first wanted to build support for the road as an endeavor for the nation as a whole, not just a project of the states and towns along the route. A committee would be appointed later to research and determine the exact route.

To fund this grand project, Fisher proposed outright donations of cash from the manufacturers of automobiles and auto accessories. He encouraged those present at the dinner party to open their company coffers and make pledges of 1 percent of their companies' gross revenues. He also proposed a five-dollar membership to be pitched to the public at large, starting with every automobile owner in the country.

Response to Fisher's appeal was immediate. Within thirty minutes of his speech he had his first major pledge: a $300,000 offer from Frank A. Seiberling of Goodyear, who made the offer without even consulting his board of directors.

Here was an idea whose time had come. The automobile was well established by 1912, but good roads were not. Some 180,000 motor vehicles were registered in the U.S. by 1910. Those vehicles plied the streets and roads during good weather but spent the winter and wet seasons up on blocks in the garage. People were beginning to look hard at the motortruck for moving goods; it was already beginning to displace horses and wagons for city deliveries. Auto drivers were becoming more adventurous as their machines became more reliable. A

few hardy souls had even driven clear across the continent by automobile, and anybody who'd been very far at the wheel was complain-
ing about the roads.

The motorists of 1912 had some two and a half million miles of
road to drive, but less than 7 percent was improved in any fashion.
Improvement usually meant some grading or graveled macadam;
oiled macadam or asphalt would come later. A few hundred miles of
brick was the extent of the hard paving, because concrete was as yet
virtually unknown as a road material. Most of the roads across the
nation were just plain dirt. A rural road could be defined as the space
left between fencerows that a farmer wasn't allowed to plow and plant.
The term was "natural" road, implying that nothing had ever been
done to improve the thing, and each car or wagon that passed did its
share to deepen the ruts. In good weather the roads were terribly
rough and dusty; in wet weather they were usually impassable.

To make matters worse, most of these roads didn't really go anywhere. There was nothing resembling a system of roads; instead most
roads extended outward from rail centers where they served to bring
the farmer into town with his goods. They resembled the spokes of a
wheel, radiating from towns of importance, with no attempt to connect the roads of one area with another. The farmer did the best he
could between the farm and the county seat, but when he wanted to
go to the next town down the line, he rode the train.

Road building and maintenance were entirely the province of local
government, and "working off the road tax" was a term that meant
leaning idly on the handle of a shovel. There were no federal funds
for roads in those years, and the tiny state and county appropriations
were said to be wasted on "pork barrel" projects that had little effect
on the condition of roads. "The highways of America," Fisher wrote to
a friend, "are built chiefly of politics, whereas the proper material is
crushed rock or concrete." In fact, twenty states had no road departments of any sort.

The long-distance motorist of 1912 was a pathfinder in every sense
of the word. Gulf Oil wouldn't invent the free gas-station road map
until the following year, and guides existed only for the eastern states.
Stopping for local advice was of limited help. Motorists found that
they had to stop every few miles to ask directions; a farmer or resident seldom had any knowledge of the roads beyond a fifteen-mile
radius.

Bellamy Partridge, contemplating a transcontinental trip about that
time, queried A. L. Westgard, the pathfinder for the American Automobile Association, for advice. Says Partridge:

*He wrote that he thought I would find the trip a very enjoyable adventure
if I was fond of changing tires and digging a car out of sand pits and mud
holes. He warned me, however, that west of Omaha I would find the roads
very informal, signboards about 1,000 miles apart, water as scarce as gasoline, and no bridges across streams under 50 feet in width. He also called
my attention to the fact that there was no such thing as a road map for the
country west of Chicago.*

Carl Fisher was not the first to propose a transcontinental highway
(the AAA had suggested one as early as 1902), but he was the first to
propose a road with a real improvement plan and the means to fi-
nance it. Other routes were drawn on some maps, but they were for
the most part unmarked, unimproved, and of scant transportation
value. A few months prior to Fisher's announcement of his Coast-to-
Coast Rock Highway, a group of promoters had left Los Angeles in
a Locomobile headed east to plot a route between the coasts. They
sought to raise money from towns along the way for basic improvement.

But Fisher had bigger ideas: he had his sights set on targets where
money for a real highway could be raised. Late in October of 1912,
Fisher published an *Ocean-to-Ocean Highway Bulletin,* a publication
listing contributors and intended to generate enthusiasm for the project.

On the cover Fisher had placed a tantalizing map of the United States depicting several of the phantom trails of mud that crossed the country—the Northwest Trail, the Sunset Trail, and others—but had intentionally drawn no clues as to where his grand highway might run.

With the enthusiasm of the Indianapolis automobile manufacturers behind him, Fisher began his campaign in earnest. He wrote letters, sent telegrams, and paid personal visits to representatives of the automobile trades across the United States. It was clear to Fisher and to the men of the infant auto industry that the success of the motorcar lay not only in better automobiles but in better roads as well. For the first time, manufacturers found that they could produce more cars than the market could absorb. The public, while ever more excited about the new-model cars, was seeing a real limit to the usefulness of even the best machines on roads that were clouds of dust in dry weather and mud when it rained. If the public were to see a clear trend toward highway improvement, it might stimulate sales and advance the cause of the automobile everywhere.

Within thirty days of the announcement of the Coast-to-Coast Rock Highway, Fisher had a million dollars in pledges and considerable ink in the nation's press.

Transcontinental fever was growing in the public imagination. On the day that details of the Fisher plan appeared in the *New York Times,* A. L. Westgard was crossing Nebraska eastbound on the old Overland Trail, later to become the Lincoln Highway. He was making the second of three back-to-back transcontinental pathfinding trips for the AAA to lay out new routes across the country. At the same time, the Federation of American Motor Cyclists was making preparations for a cross-continent cycle relay with hopes of setting a record for carrying a message from one coast to the other, and an Alco truck, carrying the first coast-to-coast truck merchandise, had reached Reno heading west.

Industry pledges and smaller donations increased as word of the new highway spread during the early months of the campaign. Newspapers and magazines wrote Fisher asking for articles. New-car orders were even being conditioned upon completion of the highway in time for the exposition. Communities across the country wondered if they would be selected to be on the route.

Carl Fisher knew that the support of Henry Ford was essential to the success of the ten-million-dollar road fund. By 1912, the 118,000 Model Ts already on the road had made the Ford Motor Company the largest in the country. Three quarters of the cars then in use were Fords. Henry Ford was already an American hero, the young mechanic from Michigan who, with limitless drive and keen automotive sense, had thrown his weight behind his ideas and risen to the top of the manufacturing world. Ford was also idiosyncratic and stubborn to the extreme, and inclined to resist anything that the rest of the automobile industry thought was a good idea.

Ford millions could build a fair portion of the highway, but more important would be his goodwill toward this new project. Many manufacturers were waiting to see what Ford would do before throwing their support behind the Fisher road. If Henry Ford thought it was a good idea, then the whole world would stand behind it. Fisher fanned the flames of national enthusiasm and laid his groundwork carefully before approaching the cantankerous automaker.

Fisher made his move early in the campaign, and a reply was quick in coming. James Couzens, secretary and treasurer of the Ford company, wrote expressing the attitude of Mr. Ford:

Frankly the writer is not very favorably disposed to the plan, because as long as private interests are willing to build good roads for the general public, the general public will not be very much interested in building good roads for itself. I believe in spending money to educate the public to the necessity of building good roads, and let everybody contribute their share in proper taxes.

Fisher was not daunted by the negative reply, and he moved in with reinforcements. He wrote President Taft and asked for his help persuading Ford. Fisher also called on Vice-President Charles Fairbanks, Ford's friend Thomas Edison, and other leaders of government and business to assist in the cause.

Despite the encouragement of many people who were close associates and friends of Ford, plus letters, telegrams, and a personal visit from Fisher, Henry Ford held his ground. No, he could not support such a project. The highways of America should be built at the taxpayers' expense. Neither the Ford Motor Company nor Henry Ford was ever to allow a contribution to Fisher's highway project, although Edsel Ford later served on the board of the Lincoln Highway Association and contributed heavily of his own time and money.

It must have been clear to Fisher that the Ford refusal put the ten-million-dollar fund drive in serious jeopardy. There simply wouldn't be money or time to build the Coast-to-Coast Rock Highway soon enough to take autoists to the Panama-Pacific Exposition. But Carl Fisher would not give up. The country was now excited about this dream road, and for the first time many people imagined themselves driving their own automobiles from one coast to the other. Contributions, though small, continued to arrive. Another way needed to be found to make this highway a reality.

Early in December of 1912, Fisher received a note that was to have a great effect on the Coast-to-Coast Rock Highway. The letter offered a $150,000 pledge to the Fisher project, but more importantly, it contained an idea that was to engrave this highway in the public imagination for years to come. The letter was from Henry B. Joy, president of Packard Motor Car Company.

"I think your Good Roads Committee, who is working up the ten million dollar fund, ought to get up a protest to Congress on the expenditure of $1,700,000 in a monument in Washington to Abraham Lincoln." Instead of the marble monument that Congress was plan-

ning, Joy wanted to see Lincoln memorialized for ". . . the good of all the people in good roads. Let good roads be built in the name of Lincoln." A year old when Lincoln was shot, Henry Joy looked to Abraham Lincoln as a larger-than-life hero. Joy's father had known Lincoln and had been a great supporter of his campaign and ideals. The honorable ghost of the martyred president had hovered close during Henry Joy's childhood, instilling in him the admiration that his father had felt.

The last sentence of Joy's note caught Fisher's attention. He had been looking for a better name for this new highway. The Coast-to-Coast Rock Highway could hardly be called inspiring. He had toyed with the idea of the Ocean-to-Ocean Highway, but that, too, was pretty dull. A friend in Congress suggested that the plan would be enhanced by including some patriotic appeal in the title or routing. Someone had suggested the American Road, and that wasn't right. But Lincoln—that might just be something. He was the great hero to Americans north of the Mason-Dixon line, and the name would have profound patriotic appeal.

At this time there was another road, a proposed road like Fisher's, that was to be named after Abraham Lincoln. A year earlier a good-roads group in the East had proposed a Lincoln Memorial Road to run from Gettysburg to Washington. It was to be a wide, paved boulevard connecting the two places for which Lincoln was best known. But Congress approved and funded the Lincoln Memorial in Washington instead of the road, and the group languished. Though Joy doesn't mention it in his letter to Fisher, he probably knew of the failed plan. When later approached by the Fisher group, the Lincoln Memorial Road people gladly relinquished the name.

Carl Fisher was quick to grasp the value of the name Lincoln, and equally quick to approach Joy in hopes of involving him directly in the cause.

Henry Joy made a business out of being careful. As president of

Henry B. Joy.
Courtesy of the University of Michigan.

Packard, the responsibility of weighty decisions rested squarely on his shoulders. After a long, careful look at the books of the Fisher organization he wrote to a friend: "Personally, I am inclined to the opinion that the enterprise has strewn in front of it almost insurmountable difficulties. Yet, on account of so much work already having been put into the matter I feel that everything possible should be done to push it on." Joy's skepticism was short-lived. He soon absorbed the Fisher spirit, and by spring of 1913 became the spokesman and spirit of the highway.

Fisher's enlistment of Joy was a wise move. Detroit was fast becoming the center of the auto industry, and Joy was among the city's top capitalists. He was well respected in the industry as the head of a company that manufactured expensive cars that were the envy of many. Fisher must have also known that Joy's connections with the monied auto men of that city were better than his own. The careful and deliberate Joy provided a fitting complement to the promoter Fisher. Both were highly motivated and idealistic men, but Joy provided a stabilizing influence that the organization sorely needed in the face of the Ford refusal. Both men must have realized that the completion of a rock highway by 1915 was in serious doubt.

In the early months of 1913, contributions were coming in much more slowly than they had been after the initial splash of publicity, much too slowly to complete Fisher's dream anytime soon. Public interest was beginning to slow because of the lack of action. Was this highway project going to be nothing but a burst of hot air? A way needed to be found to excite public interest.

There was still considerable interest in the project, but mostly because the nation was crying ever louder to learn the intended route of the road. Though rumors were rampant among automobile interests and the people of towns across the country, Fisher had leaked no details of possible routes. He wished this road to be a proud ribbon for the entire nation, not just for those communities it passed through.

He wanted to eliminate regional squabbles and to enlist support and enthusiasm from people all over the country. In reality, no conclusions had yet been reached about the best route, but Fisher understood the art of promotion and knew that if he pretended to keep the intended route a secret, the press and public would keep the issue alive.

During the spring of 1913, Fisher hurriedly called several informal and closed meetings in Detroit, where he assembled several men who would soon be the power behind this highway. He had selected them carefully and had seen to it that the Fisher persuasiveness had won them over completely. Fisher reported some four million dollars in subscriptions for the highway, including substantial new offers from the Hudson Motor Car Company and Willys-Overland. The assembled men knew of the Ford refusal, but all remained optimistic, or at least hopeful, that the money could be raised. Fisher's enthusiasm pushed aside any hint of misgivings concerning the road fund. Smaller amounts were coming in from across the country: Seattle, Denver, Texas, Massachusetts.

Fisher also presented a preliminary prospectus detailing substantial information about possible routes. Among the details of road conditions and feasibility, the prospectus stated that whatever route was selected to become the first transcontinental highway it would need major improvement over at least two thousand miles of its length in order to be considered a highway in any sense of the word. In the face of an enormous task and a faltering plan for raising construction money, Fisher pushed ahead, undaunted, with enthusiasm undiminished, convinced that the job could be done. The Fisher charm won over some of the most calculating automobile capitalists of Detroit and convinced them that if they stood behind the project, the highway could be built, if not by the time of the Panama-Pacific Exposition, then soon after.

On July 1, 1913, the group met again, this time in room 2115 of the Dime Savings Bank in Detroit, where they acted to put the organization into official existence. Their small suite of offices became the headquarters for the new association. Present were Henry Joy; Roy Chapin, president of Hudson Motor Car Company; Emory W. Clark, president of the First National Bank of Detroit; Arthur Pardington, an experienced manager and long-time friend of Fisher; and Henry E. Bodman, Joy's attorney. Carl Fisher was absent.

They decided on the name Lincoln Highway for their yet unrouted road and swiftly completed the legal formalities of incorporation as the Lincoln Highway Association. They then elected the first slate of officers. Joy became president of the new organization, Fisher and Pardington vice-presidents; Clark was elected treasurer and was to hold that office throughout the span of activity. Pardington, who was also elected secretary, was placed on the payroll and administered the day-to-day operation of the association. The officers took their new task to heart and proclaimed their purpose: "To procure the establishment of a continuous improved highway from the Atlantic to the Pacific, open to lawful traffic of all description without toll charges: such highway to be known, in memory of Abraham Lincoln, as 'The Lincoln Highway.'"

All they needed now was a route.

It seems somehow characteristic of Fisher that he didn't attend that first meeting of his Lincoln Highway Association. He was the catalyst, the spark plug, the idea man. The details could be left for others to complete—he had to keep moving. At the hour the new directors were signing the articles of incorporation, Fisher was leaving Indianapolis with the Indiana Automobile Manufacturers' "Hoosier Tour" to the West Coast. This tour was intended as a reliability run for Indiana-manufactured autos, a trip to demonstrate the feasibility of long-distance auto touring in the West, and as an examination of possible routes for the Lincoln Highway.

It is also quite possible that there was bad blood between Fisher and Joy. Both strong-minded men, they had clashed bitterly earlier in the year about how the name Lincoln was to be applied to the organization—they agreed on *Lincoln* but disagreed vehemently on the exact use. What Fisher wanted to call it and what Joy wanted to call it is unknown, but it took months of pussyfooting back and forth by Pardington to resolve the issue. No doubt there were other issues they disagreed upon, and the untimely departure of the Hoosier Tour might have meant a timely exit for Fisher.

They probably also clashed about the Hoosier Tour, which was to cross Missouri, Kansas, and Colorado on the way to the West Coast. It would cross Berthoud Pass, elevation 11,314 feet, high even for the lofty state of Colorado, and travel through narrow canyons and across desolate plateaus between Denver and Salt Lake. Joy was very familiar with the roads of the West, having driven them often even before his association with the Lincoln Highway. He knew this to be a very poor choice of routes. Early transcontinental trips had been made successfully through the passes of Colorado, but the drivers had found the going very rough. Joy had been president of Packard in 1903 when the company sponsored Tom Fetch's transcontinental trip—the second such auto trip in American history. Fetch made his trip over much of this route and wished he'd taken the northern route across Wyoming. Joy knew that by following a route across Nebraska and through southern Wyoming a motorist could reach the high point across the mountains at less than nine thousand feet, over much easier terrain.

Although Fisher took some pains to make a distinction between the route of this tour and the intended route of the Lincoln Highway, he must have imagined that the highway would pass through the states of Colorado and Kansas, and all but promised the governors and the people of those states that it would. Fisher wrote back to the directors of the new association, "The rivalry that exists between the different sections of the States through which this road is proposed, is going to be quite a factor to straighten out, but I believe it can be done." Fisher was under considerable pressure to decree the highway route to be the same as the route of the Hoosier Tour, what with banquets, speeches, free gasoline, promises of great road improvement, and strong handshakes from Governors Hodges of Kansas and Ammons of Colorado. Carl Fisher even supported the formation of Lincoln Highway booster clubs in the towns they passed through. How enthusiastic would people have been for this highway if they didn't believe that it would pass through their towns? By this time the pressure to select the route was so great that almost everyone, including Carl Fisher, believed or wanted to believe that the Hoosier Tour was laying out the Lincoln Highway. In fact the Lincoln Highway was destined not to pass through Colorado or Kansas at all.

The men who now formed the association had determined three important factors that would govern the choice of the route: first, the directness between New York and San Francisco; second, the proximity of population centers and points of scenic interest; third, the "amount and character" of support afforded the association by communities along the way. Fisher saw gain for the highway by pitting states against states and possible routes against one another to see who would produce the best promises about future road improvement should the highway pass their doors. With some delight he noted ever-inflated promises among the people and communities along the route of the Hoosier Tour.

Directness was all that mattered to Joy. With an ideal not unlike that of the interstate highway planners forty-five years later, he sought a route that would take advantage of easy terrain and natural paths, and one that for the most part would bypass scenic attractions and larger cities, avoiding congestion and narrow roadways. Connecting roads easily could be built to Yellowstone and Chicago; what was needed was an open, unobstructed pathway from coast to coast.

The Hoosier Tour of 1913 did little for the Lincoln Highway other than create confusion about the intended route and set the stage for misunderstanding. When Fisher returned later that summer, it was decided that all else must be put aside and the route selected and announced. Rumors flashed across the country, were picked up and echoed by the press, until everyone had a different opinion as to where the road would go. While Fisher believed that his road was to be for all Americans, his tactics had only fueled the flames of sectionalism. Many donations were being made conditional on the road passing through the donor's state, or past his front door. The hope of a road for all America was fading as towns and states clamored to be put on the highway and endlessly queried the association to ascertain the intended route.

The officers decided that they would complete their route selection process and make their case before the annual Conference of Governors to be held in Colorado Springs in August. This was to be a plea for unity. The Lincoln Highway Association wanted universal support for its highway, but knew that disappointment on the part of states not touched by the road was inevitable.

Armed with everything they could find, the directors of the association set about making their decision. They assembled all available guides and maps, and the volumes of notes kept by Joy on as many as ten cross-country trips over various routes. So little published road information was available that the group often had to rely on railroad timetables to provide mileages between cities. After evaluating the road condition of many routes, figuring many possible mileages, and listening to much firsthand experience, and, after what must have

been considerable argument and deliberation, the route was chosen. The directors drew a map of their route and wrote a speech for the governors. Joy, Fisher, and Pardington left Detroit by train for Colorado Springs. Their speech was called "An Appeal to Patriots." It was Joy who stood before the governors late on the afternoon of Tuesday, August 26, 1913, at a secretive meeting held in a room at the Antlers Hotel; the public and the press had been excluded.

The appeals of sections have been heard. The arguments of all interests have been and are being weighed. Shall the Lincoln Way be marked on the map from large city to large city? Shall it be from point of interest to point of interest? Shall it be a highway from New York to San Francisco, as direct as practicable considering the limitations by Nature in the topography of the country?

For decision, the hopeless divergence of conflicting interests and opinions must be eliminated, and the practical conditions only must be considered. The decision must be confined to one permanent road across the country to be constructed first, no matter how desirable others may be and actually are. Such has become the basic principle guiding the Lincoln Highway Association. It is seeking to decide wisely a matter which must be decided right in order to eliminate the petty hauling and pulling and opposition which would be fatal to the great patriotic work, and which would thus postpone beyond our vision so laudable a project.

It seems to us but yesterday that the Panama Canal was begun, and yet almost tomorrow it will be open to the world.

Henry Joy then presented the map and described the chosen route of the Lincoln Highway.

2

A Highway for Lincoln

It is a name to conjure with. It calls to the heroic. It enrolls a mighty panorama of fields and woodlands: of humble cabins and triumphant farm homes and cattle on a thousand hills: burrowing mines and smoking factories: winding brooks, commerce-laden rivers and horizon-lost oceans. And because it binds together all these wonders and sweeps forward till it touches the end of the earth and the beginning of the sea it is to be named the Lincoln Highway.

— Reverend Frank G. Brainard
of Salt Lake City, preaching
from the pulpit, Sunday,
October 19, 1913

The Lincoln Highway began at Times Square, Broadway and Forty-second, New York City, and ended at the Pacific Ocean in Lincoln Park, San Francisco, 3,389 miles west.

From New York, it angled southwest through Trenton and Philadelphia, where it turned west. The highway crossed Pennsylvania and the ribbed Alleghenies, passed through Pittsburgh and emerged in Ohio and the Midwest near Canton. It tied the cities of Van Wert, Fort Wayne, and South Bend, and closely skirted Chicago to the south and west. From Chicago it pointed directly west across Illinois and Iowa, finally turning southwest to cross the Missouri River at Omaha and to make a connection with the Platte River route. Joy's map showed the road passing through Cheyenne and across the broad uplands of Wyoming, through Echo Canyon to Salt Lake City. It followed the edge of the Great Salt Lake desert around the south end of the lake, then linked the remote Nevada towns of Ely, Eureka, and Austin with Reno. The main route crossed the Sierra Nevada over Donner Pass and made a long descent to Sacramento, then looped around to the south through Stockton to enter San Francisco via Oakland and the bay ferries.

The Lincoln Highway led to the Panama-Pacific Exposition but officially ended in Lincoln Park, overlooking the Pacific Ocean. It ran through or touched twelve states: New York, New Jersey, Pennsylvania, Ohio, Indiana, Illinois, Iowa, Nebraska, Wyoming, Utah, Nevada, and California. It did not cross either Kansas or Colorado. This was clearly Joy's highway.

Reaction at the governors' conference was immediate. The warm welcome extended to Joy, Fisher, and Pardington by the citizens and governor of Colorado vanished. Governors Ammons and Hodges of Colorado and Kansas were bitterly disappointed and no doubt embarrassed before the people of their states for having so heartily supported the Indiana tour, for being led to make great promises in support of a highway which was now to bypass them. Considerable angry persuasion was brought to bear in hopes of changing the route to favor the forgotten states. The strident appeal of Ammons and the polite courtesy owed the host state of Colorado pressured the highway men to rethink their plan. After considerable deliberation, they agreed to a dogleg of the highway from Big Springs, Nebraska, southwest to Denver, and thence due north to rejoin the main Lincoln Highway at Cheyenne. Though it calmed Ammons, the dogleg opened Pandora's box to anyone who sought to bend the Lincoln Highway in any direction for any reason. What was intended to solve the problems created by the Hoosier Tour in turn created a nest of greater problems.

Inexplicably, the association map also showed an alternate and longer route between western Nevada and Sacramento: it passed through the capital city of Carson City, crossed Echo Summit, and descended through Placerville to Sacramento. It was intended as a scenic route and perhaps was established to appease some state interests. This route, referred to as the Pioneer Branch, was never explained, and in fact the association often ignored it, but it stayed on association maps and guides until the highway numbering system came into general use.

The public announcement of the highway's route followed on September 14, 1913, marking the culmination of a huge publicity effort on the part of Secretary Pardington in Detroit. Newspapers across the land printed the "Appeal to Patriots," the Proclamation of the Lincoln Highway Association, a list of towns and cities along the route, and the map, including the Denver dogleg.

The Lincoln Highway Association set aside October 31 as a national day of celebration for this memorial to Lincoln. The farmers of Indiana marked the occasion by combining it with Halloween and set jack-o'-lanterns on fenceposts for many miles along the route. Parades, bonfires, good-roads speeches, dances, auto events, and fireworks in towns along the route across the country fired the imagina-tions of many who now took up the Lincoln Highway craze. Clergy were encouraged to speak of Lincoln in their sermons, and perhaps to mention the new highway named for him.

The association created a membership program by which for five dollars the donor would get a certificate, a membership card, and an enamel radiator emblem. President Woodrow Wilson sent in his five dollars and received membership certificate number one. Civil War veterans who had served under Lincoln sent in their contributions. Ministers, lawyers, bankers, and farmers, some who didn't even own automobiles, sent letters and pledges. By the end of 1913, there were contributions from people in forty-five states and several foreign countries. These donations actually amounted to little real capital, but the association milked them for all possible publicity value.

The directors created a system of local "consuls"—ambassadors really—who were picked to represent the association in places along the highway. They appointed a consul for each state first, then asked these men to seek county and town consuls. The consuls were merchants, attorneys, and editors who represented the association in local and regional affairs, kept headquarters informed of events and decisions that affected Lincoln Highway matters, provided tourist information and assistance, and were usually the first to determine the condition of the road after bad weather. At maximum, there were nearly three hundred consuls scattered across the country.

The Lincoln Highway Association was soon flooded with letters from people and communities trying to get the route changed. Road-minded citizens across the country eagerly scrutinized the route and looked for ways to make it straighter, more direct, and, most important of all, to make it go through their own towns. Since an exception had been made for Denver, why not one for other towns? Each inquiry was politely answered, in hopes of minimizing the anger and disappointment many bypassed communities felt. As committees were being formed in towns and cities astride the route to promote and help build the road, committees were being formed in towns off the route in hopes of getting the route changed.

Unlike many other states where geography channeled travel along river valleys, around mountain ranges, or through passes, Iowa had several possible, nearly parallel routes that could have been selected, and here the desire to twist the road to a new path was strongest. Civic leaders in towns on the selected route nodded and complimented the Lincoln Highway people for making the obvious and best route selection, while those in towns on routes not selected cried out at the ignorance and blindness of the Lincoln Highway Association for overlooking them. The ink was barely dry on the map when towns and citizens began scheming ways to get the route changed to include them. What a proud thing for a city, what an opportunity to show travelers from far and near just how wonderful the town was! Little burgs across the country dreamed of making a place on the map at last.

Boone, Iowa, congratulated itself for being on the selected route for the Lincoln Highway but kept a close eye on the doings of several towns intent on changing the route and having the road for themselves. The *Boone News Republican,* under the headline "FORT DODGE, DUBUQUE AND OTHER TOWNS SORE," warned the good-roads men of Boone that "they cannot go to sleep and hope to retain possession of the highway that is destined to capture a goodly portion of the cross-country automobile travel of the country."

Three weeks later the paper warned of increasing pressure and organizational efforts on the part of communities desiring the route and quoted a piece from the *Perry Chief* as ample evidence that people were hard at work trying to divert the highway: "The opportunity which has come to Perry for the construction of a 'loop' in the famous Lincoln Highway is one which should not be overlooked. . . . Perry should be on the map in the Lincoln Highway. It can be placed there

if the people of the community get behind the committee which has the matter in charge and will shove."

About this time Joy received a letter from an indignant Iowan who claimed that his route across the state was not only better but a good deal shorter as well, and demanded that the Lincoln Highway be rerouted through his town. The patient Joy wrote back and agreed that, yes, his route was shorter, but only because Iowa was narrower where this route crossed; the difference would have to be made up in Illinois.

The association, under Joy's firm command, refused to budge, refused to make the slightest deviation in the highway's path. It was learning its lesson from the loop to Denver. Two years later, in 1915, that jog was quietly dropped from association maps and guides. Once again, Colorado was furious and citizens of that state would not give up their route. As a result, the 1916 guide published by the Lincoln Highway Association warned the traveler at Big Springs: "The tourist wishing to follow the official Lincoln Highway straight west through Cheyenne, Wyoming, should not be diverted at this point by markers or signs indicating that the Lincoln Highway turns southwest here to Denver. Numerous markers have been placed here to mislead the tourist." Few people outside Colorado noticed, but within that state little good was ever said about the Lincoln Highway.

That people across the country should think it easy to bend the highway to one path or another was really quite logical. As yet this great road was nothing more than a line on Henry Joy's map; nothing had changed just because the name Lincoln had been applied to this loose collection of more or less end-to-end roads. They had been ordinary dirt roads one day and dirt roads called the Lincoln Highway the next.

But the highway was drawn as a grand boulevard like the Pennsylvania Railroad—the "Standard Railroad of the World"—or like the

Times Square, New York City, starting point for the Lincoln Highway.

On our way up Fifth Avenue, two or three times in the traffic stops, we found the motors of friends next to us. Seeing our quantity of luggage, each asked: "Where are you going?"

Very importantly we answered: "To San Francisco!"

"No, really, where are you going?"

"SAN-FRAN-CIS-CO!!!" we called back. But not one of them believed us.
—Emily Post, 1915
By Motor to the Golden Gate

Merchants and Drovers Tavern, established about 1750 along the King's Highway at Rahway, New Jersey.

King's Highway, later the Lincoln Highway, Kingston, New Jersey.

I can't remember at just what point we began talking about driving across the continent. I'm sure the idea was mine originally, for I was always reading Walt Whitman, Bliss Carmen, Vachel Lindsay and all that company of vagabonds who sang nostalgically of the open road and far horizons. It was Kit, on the other hand, who got the maps and did the figuring on backs of envelopes about the number of miles per gallon. It was she who said, "Let's do it. Let's start May 31."
—Beth O'Shea, 1920
    A Long Way from Boston

Lancaster Pike marker, Paradise, Pennsylvania.

Mrs. Joseph DeHart now sells cut flowers at her home in Lancaster, Pennsylvania, where Lincoln Highway travelers once stayed the night.

Susquehanna River crossing, Columbia, Pennsylvania. Since 1913 three bridges have carried the Lincoln Highway across the river here. This twenty-eight-arch concrete bridge, completed in 1930, replaced an earlier steel-truss bridge that stood on the now-vacant stone piers. Today U.S. 30 crosses a half mile upstream on an unadorned four-lane steel span.

Town square, New Oxford, Pennsylvania.

NEW OXFORD

N.Y.   S.F.

199   3132

Pop. 1,000. Adams County. Two hotels, $1.50–$2.00, American. Two garages. Local speed limit, 12 miles per hour, enforced. Route marked through city and county, signs at city limits. Two railroad crossings at grade, protected. One bank, 14 business places, 1 express company, 1 telephone company, 1 newspaper, 1 public school.
—Complete Official Road Guide of the
  Lincoln Highway, 1916

1928 Lincoln Highway marker along bypassed route, Fayetteville, Pennsylvania.

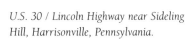

U.S. 30 / Lincoln Highway near Sideling Hill, Harrisonville, Pennsylvania.

On the map, it appeared an easy day's run—some thirty miles shorter than our first day's. In our calculations we overlooked the considerable matter of the Alleghenies.

All day long we boiled and wheezed up them and scorched our brake bands sliding down their further slopes, ascending and descending with the regularity of an office building elevator. At nightfall, despite our most frantic effort, we had logged only one hundred and forty-five miles, and crept, late and dispirited, into camp at Greensburg.
—Frederic Van de Water, 1927
   The Family Flivvers to Frisco

Tourist cabins, Breezewood, Pennsylvania.

Lincoln Highway Farm, Schellsburg, Pennsylvania.

We were now to traverse the Lincoln Highway and were to be guided by the red, white, and blue marks; sometimes painted on telephone poles, sometimes put up by way of advertisement over garage doors or swinging on hotel signboards; sometimes painted on little stakes, like croquet goals, scattered along over the great spaces of the desert. We learned to love the red, white, and blue, and the familiar big L which told us that we were on the right road.
—Effie Gladding, 1914
   Across the Continent by the Lincoln Highway

Old Pennsylvania Road to Pittsburgh, later the Lincoln Highway, near the summit of the Alleghenies, Somerset County, Pennsylvania.

*George Westinghouse Memorial Bridge,
constructed in 1930 on the Lincoln
Highway, East Pittsburgh, Pennsylvania.*

U.S. 30, Ohio River crossing, East Liverpool, Ohio.

East Liverpool, Ohio.

Lincoln Highway, Dalton, Ohio.

"Can you tell me," I asked her, "which is the best road to California?"
    Without hesitating she answered, "The Union Pacific."
—Emily Post, 1915

Oregon Trail, steeped in importance and history, where thousands had crossed to the Golden West. It was drawn boldly to signal the bold intentions of the association. On the map it looked like the Appian Way, having only the most gentle turns or perturbations; it looked to be a smooth, gently curving boulevard spanning the continent. Following the line with the eye, the motorist could imagine rolling across the country, with one hand draped over the wheel, slowing only for larger towns and stopping only for a dozen or so main intersecting roads.

Everybody knew that this road, like almost every other road in the country, was terrible. Some people—no doubt residents of towns left off the route—called it a red line on a map connecting all the worst mudholes in the country. And to a great extent they were right. As drawn, it overlooked the hairpins of Donner Pass, the endless zigzags along the Platte River in Nebraska, the bottomless gumbo of Iowa, and the grades of the Alleghenies. West of Pittsburgh it lacked any resemblance to a connected route, there wasn't a single foot of permanent paving outside of a few towns and cities, and there was but 650 miles of crushed-rock surfacing out of some 3,300 miles of road. In places out West, the new Lincoln Highway grew faint as wheel tracks fanned out across the desert, or disappeared altogether beneath the currents of streams in flood.

This fat line was really meant to fool no one. The highway was depicted not as an existing road but as a road to be. It was drawn as an ideal, a line to prod the imagination and bankbook into action. Not a shovelful of earth had yet been moved to build this year-old dream—thus far, the Lincoln Highway had been the work of only a few promoters, an automobile adventurer or two, and a draftsman who drew a line on a map—but despite the complaints of those left off the route, this highway had taken hold of the public imagination. Fisher and Joy knew that more than stone and concrete and steel

would be needed to build this highway; if America could imagine a broad boulevard spanning the continent, then somehow it could be financed, and it could be built.

But by early 1914 it was apparent to the directors that the association was in financial difficulty. Fisher's optimistic ten-million-dollar road fund essentially was dead. The amount pledged stood somewhere near the halfway mark, but after Ford's refusal to participate, and a strong dose of disinterest on the part of potential donors who found themselves left off the route, it was clear that the plan would go nowhere. None of the money pledged for construction was on deposit with the association; Fisher had honorably arranged it so that none was due until the entire ten million dollars had been pledged. It was obvious that the Lincoln Highway Association wouldn't be building this road single-handedly.

Operating capital was also a nagging problem. People along the route who sent in five-dollar memberships expected great improvement for their contributions and wanted the money spent on the road near their town. In fact, membership certificates at five dollars each hardly paid the costs of distributing them. An editor of a small Nevada newspaper wrote the association in 1914 and demanded, "What have you done with the money?" Secretary Pardington patiently wrote back that the $265 total from the state of Nevada was being used to educate those in the East about the West in hopes that touring would increase in western states, bringing much-needed revenue and expanding the tax base. The answer must have satisfied the newspaperman, for he later became a local consul for the association.

The early operations of the association had been costly; some seventy-five thousand pieces of mail were sent in September 1913 when the route was announced. Joy, Fisher, and the other founders personally underwrote these preliminary expenses to the tune of $22,000. With no money for construction and little for operation, and with great expectations on the part of the public about this great boulevard, the men of the association realized they were in a tight spot.

Joy found the key he sought in the letter from Couzens explaining Ford's refusal to participate in the ten-million-dollar road fund. If we can't raise the money directly and build this road ourselves, Joy thought, why not do just as Couzens says and educate the public to the necessity of building good roads? As public opinion changes about good highways, government will be pressured into building them as a resource for all. The country will soon see the value of this and all good highways to tourists, farmers, national defense, and the local and national economy. We will help the nation build this highway as an object lesson, as an example to everyone of the value of good all-weather highways.

The ten-million-dollar road fund was quietly abandoned and a new plan was devised. First, the association would encourage the marking of the entire route; second, it would petition cities, towns, and counties to rename local parts of the route Lincolnway.

And lastly, Joy proposed to publicly abandon the goal of building a gravel road at association expense and completing it in time for the fair, and instead to concentrate on assisting the country in building its own road. The Lincoln Highway Association would continue to seek donations for construction but would leave the burden of major funding to the states, counties, and communities along the way. It further proposed that this be a model road, a road as permanent and enduring as any could be—from coast to coast, it would be built of poured concrete. It would take much longer than the original Fisher plan of using gravel, but it would be a truly lasting monument to Lincoln.

As the first step in the process of educating the country about concrete roads, Joy proposed that the association fund and oversee the construction of demonstration miles in places where improvement was most needed. He dubbed them "seedling miles" in the hope that they would germinate larger contributions for road building and sprout tendrils of additional concrete. Joy and the association proposed them for Illinois, Iowa, Nebraska, Wyoming, Utah, and Nevada. Said Joy:

It seems clear to me that I would prefer to pursue the education of the public and the collection of funds in a small way until the public mind is more saturated with the objects and intent of the Lincoln Highway Association, and until the times may seem more propitious for asking large contributions.

In groping about in my own mind for some concrete, tangible thing to do which would place the Lincoln Highway more plainly before the communities, I have always had in mind building in each one of the states from Illinois to Nevada inclusive, a sample mile of the roadway. . . . The work of the Lincoln Highway Association would then become a tangible, feasible, sensible expenditure of the funds being raised.

Now this was an idea to raise eyebrows of even the daring among highway promoters. Many road officials across the country considered concrete to be experimental at best. That some highway association should propose a road of concrete stretching better than three thousand miles clear across the country was certainly bold, and many must have thought it foolhardy. The first rural concrete road had been poured only five years earlier in 1908. This mile-long stretch in Wayne County, Michigan, drew motorists from miles around who came to drive the road and marvel at its smooth, even surface. They knew it couldn't last. It didn't quickly show signs of wear, but moisture and Michigan winters were sure to be its undoing. How could this stuff pour like liquid one day, then, a few days later, be rock hard, impervious to pounding wheels, horses' hooves, freezing, and thawing? No doubt many pocket knives came out to scratch at the hard surface, as people shook their heads and looked closely for the first signs of certain failure.

Undoubtedly Henry Joy was one of the motorists who had come to see this experimental mile. It was, after all, only a few miles from his office. He must have stood and imagined this stuff spread all across the country.

Fisher and Joy knew that the mile of concrete on Woodward Avenue would last longer than roads made of any other material. They also had in hand a donation of a million and a half barrels of concrete from the cement industry, with the possibility of double that amount. A. Y. Gowen of the Lehigh Portland Cement Company had written Fisher on behalf of the cement industry and offered the donation, thinking the coast-to-coast highway project might be a proving ground for this new road-building material. Fisher and Joy concluded that they would use their road to educate America about good roads in general and about concrete as the best material with which to build them.

The first seedling mile was completed near De Kalb, Illinois, in the fall of 1914. With the promise of free cement from the Lincoln Highway Association, supporters there had raised two thousand dollars in cash from public donations and gathered three thousand dollars from the county supervisors. The state had prepared a grade and offered road equipment and engineering supervision. The good-roads people of De Kalb gathered their resources and requested the concrete from Secretary Pardington. He replied: "The ball is opened and you are the leader of the first dance. I have just received word from the Marquette Company that they are shipping to you 2,000 barrels of cement."

The ball had indeed begun; this was the first tangible, permanent improvement the Lincoln Highway Association undertook. It was a small but certain beginning. This slab of new pavement was only ten feet wide and was only a single mile out of at least two thousand in dire need of improvement. But it was a beginning.

The year 1914 saw the production of motor vehicles exceed the production of wagons and carriages for the first time in the United States. That year also saw the first effort at marking the Lincoln High-way. At the encouragement of the association, civic groups, business people, and general citizenry from communities along the route fanned out to paint official Lincoln Highway markers in patriotic red, white, and blue stripes on barns, trees, rocks, telephone poles, and fence-posts. What the marking job lacked in standardization or neatness it made up in enthusiasm. Anybody lucky enough to live or do business along the Lincoln Highway was proud of the fact.

The next year, 1915, saw four seedling miles built on the high-way—the second and third were opened in Nebraska and were fifteen and sixteen feet wide; another was opened in Illinois and one in Indiana.

The Lincoln Highway Association was shrewd in seeing that these miles of road were generally located in the countryside, far from town limits so as to provide the greatest contrast possible between unimproved dirt and smooth concrete. This also allowed the seedlings to grow in either direction. A few were extended from town limits in one direction or another, but to connect one end to the cobblestones of town softened the impact of this superior road surface.

The successful publicity of the Lincoln Highway project fueled a great surge of highway fever in the United States. Certainly there had been named roads before the Lincoln—the Midland Trail, Sunset Trail, Northwest Trail—but just as their names suggest, most were "trails," or primitive roads that had little means or hope of improvement. There weren't very many of them, and in most places they remained unimproved despite efforts to raise money among towns and cities along their routes. Sufficient money just couldn't be found locally to take on the mammoth job of building hundreds of miles of road. Most roads were marked, promoted, and left in their unimproved state.

After the initial splash created by the men from Detroit and Indianapolis, new roads began to spring up like weeds in an untended garden. To the outside world, it looked like the Lincoln Highway people had money to build an all-weather road and like construction starting at one coast and ending at the other would begin any day now. Few outside the association board of directors knew the truth. The country saw what was planned along the Lincoln Highway, and people everywhere had to have their own highway.

In a sense it was a direct response to the way Fisher had conducted his campaign. By keeping the intended route a secret in the early stages, he had planted a seed in all the towns from Maine to Texas that it was possible that they might be selected to be on the route. When the Lincoln bypassed them, they all turned their backs on the Fisher and Joy project and instead concentrated on finding or creating a road of their own. The Midland Trail people rushed in to fill the vacuum in Kansas and Colorado, and in other towns everywhere road associations sprung up, speeches were made, brochures were printed, and eager paint crews left to mark telephone poles with a new logo.

Some, like the Midland Trail, which ran from Washington, D.C., to Los Angeles, were intended to compete with the Lincoln. Others ran north and south or ran in no apparent direction, seeming to wander between arbitrary terminals. They connected the East Coast with the West, large cities with large cities, and small towns with places no one had ever heard of. In a few years following the establishment of the Lincoln, many named highways sprouted to lure the traveler this way or that, and most importantly through specific towns and past specific businesses.

These roads were usually the product of local boosterism; civic pride on the part of townspeople and businessmen prompted them to climb on the good-roads bandwagon, because if Perry, Iowa, or Colorado Springs, Colorado, couldn't have the Lincoln, they could have the Des Moines, Fort Dodge, Spirit Lake, and Sioux Falls Highway, or a branch of the Midland Trail. Towns worried that if they didn't establish and promote their own road, all the traffic, all the potential business, would soon pass them by on a competing route. Good-roads fever was taking hold.

Unlike the Lincoln Highway, most of these new roads didn't have the benefit of any plan or intention of real improvement. Work was usually done locally, when a citizen committee or a handful of county workers would drive a few miles to replace a culvert or washout. Farmers were hired to drag the road and smooth the ruts after rain. These efforts kept the road open but had little effect on building a better path. In addition, directness was seldom a priority—what mattered was that the road went through the town in question. Some went from place to place like ivy on a wall, seldom in a straight line and with no intention of traversing the shortest possible distance. Travelers complained that the marked road sometimes took them around three sides of a section in order to pass through some town, when the same distance could have been covered by a straight line and a single mile.

Meanwhile, at a time when national enthusiasm was growing most quickly for his dream highway, Carl Fisher stepped away from any active role in the organization. By 1915 he resided in Miami Beach, Florida. He usually returned to Detroit for board meetings, and he remained a vice-president on association letterhead until the organization disbanded, but now that the project was well under way, now that many others had picked up the Lincoln Highway spirit, his interests turned to other things. Perhaps friction between him and Joy encouraged his absence, but it is certain that Fisher, always the promoter, always the man with a new idea, had other projects in mind.

Fisher had joked about retiring, but anyone who knew him figured he had other plans. His new dream was as bold as a coast-to-coast highway had been, and just as well timed. Fisher planned to enlarge and develop an alligator-infested sandspit opposite the city of Miami into a playground for the rich with huge hotels, wide streets, and miles of beaches. He hoped to attract his old cronies from the North, the men of industry from Michigan, Indiana, and Illinois, who, weary of snow and cold, would flock to the Atlantic Ocean at Miami Beach and join him in the sun. Skeptics predicted that he would quickly go broke dredging sand from the sea to build a footing for his dream city, but his detractors underestimated the tenacity of Carl Fisher.

To cast his lure to the men of the frost belt, Fisher turned to his highway promotion experience and devised another great road—the Dixie Highway. It had several braided branches that began in northern Michigan and at Chicago, amid the grand homes of the very people he hoped to draw south. The tendrils wound their way south and east through Ohio, Indiana, Kentucky, Tennessee, and Georgia, and finally ended at Fisher's door in Miami. Like the scent trail left by sugar-seeking ants, the Dixie Highway led irresistibly from home to the sweet reward on the Atlantic coast.

To Henry Joy, the great proponent of the direct route, this wandering peavine of the Dixie must have seemed a poor excuse for a highway. It was everything that he disliked about the highways of those days. The Dixie placed little value on directness, and with multiple routes it was diffuse, dilute, and subject to all manner of political pressures. Was this the sort of road the Lincoln Highway would have become had Joy not taken the initiative and insisted on a direct, straight route? By 1915 the Lincoln had become Joy's highway; Fisher went on with other projects and was never to have much to do with it again.

3

The Open Road and the Flying Wheel

The Lincoln Highway was nearly two years old in the spring of 1915 as Emily Post and her traveling companions made their way west from New York. The Great War was raging in Europe, and President Wilson was adamant that the United States would not become involved. Production of Fords passed the one-million mark, and Oldsmobile began offering a top and windshield as standard equipment. In January of that year, Alexander Graham Bell inaugurated long-distance telephone service between New York City and San Francisco. "Hello Frisco" became the song of the year.

Along with the war in Europe, the new exposition was making the news. The Panama-Pacific International Exposition had opened in San Francisco on February 20, and Americans were flocking to the West Coast to see the latest wonders of art, science, and technology. It was a celebration of Balboa's discovery of the Pacific Ocean some four hundred years earlier and a fete for the new Panama Canal, which had opened in 1914. Most Americans who went to the exposition went by rail; the newspapers were full of railroad advertisements for special fares and excursion trains. But a handful, perhaps a few hundred, set out for the fair by automobile on the new Lincoln Highway.

In the spring of 1915, while ships were passing from the Atlantic to the Pacific through the new canal, automobiles were passing from the Atlantic to the Pacific on the Lincoln Highway.

But not Emily Post. She and her companions, scared off by the reported construction on the Lincoln in Pennsylvania, went by way of Buffalo, where they rested from the first lap of their trip at the Hotel Statler, "a commercial hotel with a much advertised and really quite faultless service that carries the idea of personal attention to guests to its highest degree," wrote Post. The evening before they resumed their westbound trip, a man stepped into the lobby who was covered from head to foot with yellow dust. Seeing Emily stare at this ocher apparition, the clerk said: "Oh, he's just a motorist who has come from

Cleveland. Gives you some idea of the roads, doesn't it?" They left the next day in an uneasy frame of mind, headed for Chicago and the Lincoln Highway.

Emily Post was a woman of New York society, a woman accustomed to society teas, debutante balls, and nearby servants. Her life so far had been conducted in the homes and watering places of the well-known well-to-do of New York and Europe. Although her most famous contribution to the literary world, *Emily Post's Etiquette,* was still seven years in the future, she had proven herself a worthy writer with the publishing of two novels. An editor and friend, Frank Crowninshield of *Collier's,* asked her to take this motor trip across the country and report on what she found.

Although she had misgivings, she recognized this as an opportunity to experience and write about something far outside of her normal, sheltered society life-style. Besides, it played to her acute sense of independence and her growing dislike for the slow, dirty trains.

She agreed to go, but with some conditions. She was not going to rough it, and she would travel only as far as it was "pleasurable." No camping, canned beans, or pathfinding. It was to be hotels and dining rooms all the way, and only the best of those. Sleeping on the ground was out; her greatest terror was of rattlesnakes. If the trip became difficult, the roads too bad, the hotels too far apart, the car simply would be put on a train and shipped the rest of the way to the West Coast. Crowninshield agreed, and since Emily had never driven a car, she pulled son Ned out of Harvard for the remainder of the semester and gave him the job of chauffeur and mechanic.

Crowninshield wanted her to describe the places she went: the cities, towns and countryside, and the people she met along the way. An expense record and daily log was to be kept and published with the articles. The series was sure to be popular with the readers because they, like many Americans, were growing restless to span the continent by automobile.

Emily also was to describe this new Lincoln Highway that connected New York City and San Francisco. Interest in transcontinental motoring had taken a sharp jump a year and a half earlier with the announcement of the highway's route and the intention of completing this fine road in time for the exposition. Reports were coming back that the road had been greatly improved—bridges rebuilt, mudholes filled, signs and markers placed, so that even the traveler with a poor sense of direction could find the way. But how much real work had been done? Crowninshield wanted Post to report about the ease of travel on this much-heralded road, and to write glowingly of its safety and advantages over other routes. Then, when she reached the West Coast, she would report on the Panama-Pacific International Exposition and a lesser exhibition, the Panama-California Exposition in San Diego.

Heaped in the Posts' car was an endless array of the wrong items for a long-distance motor tour. Here was a collection of articles much better suited to a rail tour of Europe or New England. The car was burdened with Emily's luggage: a giant steamer trunk, a collection of dressing cases, hatboxes and canvas bags, a selection of coats of various weights, three rugs, a typewriter, a medicine chest, a five-pound box of chocolates, and of course, the picnic basket with the silver tea service. Added to this were bags packed by Ned and Alice. Good sense prevailed to the degree that they also carried a pair of canvas water buckets, a block and tackle, a selection of tools, and one hundred feet of stout rope. Space in the otherwise roomy car was sufficiently cramped that Alice had to be loaded into the car and then luggage stowed around and over her.

After a few days of wrestling with it in the bulging car, the wicker picnic basket was shipped home and the notion of afternoon tea was, if not abandoned outright, at least given over to tearooms and hotel dining rooms along the way. Parcel by parcel, other articles soon followed the basket home, as more practical clothing and utensils were substituted along the way. A good deal of what had been elaborately packed and taken along proved ill-suited to the rigors of long-distance motoring across the American interior.

As the days wore on and Ned, Alice, and Emily grew accustomed to travel by auto, the car became lighter and roomier, and certain traveling habits evolved. Emily developed a useful costume and a routine that allowed her to arrive at the day's destination appearing fresh and clean despite the dusty roads. Her garb consisted of a blue silk duster over a wrap suited to the day's weather. On her head she wore a close-fitting hat, shaped like a beehive, held on with yards of blue chiffon. Big goggles with yellow lenses protected her eyes. Ned thought she looked like an enormous June bug. She must have been a little self-conscious about this getup, because Ned was under orders to stop at the edge of the town where they were to lodge for the night so she could remove the chiffon and the goggles and put on a blue lace veil. Emily felt that the veil covered the dust, mud, and sunburn of the day's drive and lent an air of dignity when she clambered out of the car (it had no doors; one had to climb over the side bathtub fashion) and entered the hotel to inquire about Mrs. Post's reservation.

Good hotels and good roads saw the party to Chicago. While there, Emily took the opportunity to mingle with members of the city's finer society. She also went shopping at Woolworth's, where she bought cheap plates, spoons, oiled paper, and a kettle. To this she added a supply of canned meats, an alcohol lamp, and a saucepan. To carry it all, she bought a breadbox; to secure it she bought a padlock, and in grand total spent only a few dollars to replace the beautiful but useless picnic basket and tea set with a provision box that was a marvel of simplicity and usefulness.

With their easy warm-up on the reasonably good roads of the East behind them, and fortified with fresh supplies and motoring savvy, Post and her family left the warm, civilized Blackstone Hotel in Chicago for the rainy prairies of Illinois and the West beyond. Finding good hotels up to Chicago had been easy, but beyond the Mississippi the worst could be expected. Though dusty in the East, it had been a wet spring in the Midwest, bringing worry about the roads to come. The adventure was to begin. Here at last was the Lincoln Highway stretching to the western sunset at the Pacific Ocean.

For Post, the much-touted Lincoln Highway was a disappointment from the start. They first picked up the grand highway some thirty miles west of Chicago. It was fine for a while; then, a few miles east of Rochelle, it disappeared in a sea of mud.

If it were called the cross continent trail *you would expect little, and be philosophical about less, but the very word "highway" suggests macadam at the least. And with such titles as "Transcontinental" and "Lincoln" put before it, you dream of a wide straight road like the Route National of France, or state roads in the East, and you wake rather unhappily to the actuality of a meandering dirt road that becomes mud half a foot deep after a day or two of rain!*

Having no tire chains, they slithered along as best they could, nearly sliding into the roadside ditches several times. Finally, upon reaching Rochelle, splattered with mud and pale as a ghost from the near upsets, Emily announced that she'd had enough. She didn't care how poor the hotel might be in this little town; she wasn't going another mile until this terrible Lincoln Highway dried. To her surprise and relief, the hotel they found had private baths. It reminded her of "a Maine summer resort."

While the New York travelers took in the sights and sampled the social life of a small Illinois town, it rained steadily and the roads grew worse. Alice and Emily went to the moving picture houses, and Ned passed the hours with the men at the local garage. Occasionally

Emily Post, son Ned, and their hand-built touring car, New York City. Emily sports a fashionable hat, but one not well suited to travel in an open car. Ned is driving; since the automobile is English, the controls are on the right.
Courtesy of Mr. and Mrs. William G. Post.

they went to the west edge of town where the brick paving ended and looked out at the mud of the wet, black Lincoln Highway leading off toward the coast.

The late spring of 1915 was one of the wettest anybody in the Midwest could remember. Emily Post and her companions had muddled right into the thick of it when they reached Rochelle, where their auto trip nearly ended, only a third of the way across the continent. The rain abated for a while, but the roads would not dry under the low clouds. Emily finally decided that her encounter with the mud of Illinois was too much like roughing it. She decided to put the car on a fast freight and ship it past this Midwestern mud district—or all the way to California if that's what it took. It required considerable coaxing by Ned and the encouragement of local citizenry, plus a good dose of sunshine, to change her mind. Finally after two days in Rochelle, Emily, Ned, and Alice repacked the car and, on May 8, 1915, fortified with a set of new tire chains, started west again.

Nineteen days after Post left Rochelle, another trio of long-distance travelers started for the West Coast and the Panama-Pacific Exposition. Henry B. Joy, A. F. Bement, and Ernie Eisenhut left Detroit on May 27 in a new Packard 1-35 touring car. Bement was the new secretary of the Lincoln Highway Association, replacing Pardington, who had died earlier that spring. Ernie Eisenhut was a top mechanic for Packard who was sent along to keep close tabs on the car.

The trip had several purposes: it was to be a shakedown for this new-model Packard and a reconnoiter of conditions along the western two-thirds of the Lincoln Highway. Upon reaching San Francisco and the Panama-Pacific International Exposition, the car, hot off the highway, was to be put on display and would be a fine promotion for both Packard and the Lincoln Highway.

The Packard and its crew were to get a full test. During the three weeks since Post had struggled through, it had continued to rain across the Midwest. Travel conditions, bad enough when the Post party crossed, deteriorated markedly across Illinois, Iowa, and Nebraska. The constant rains hampered farm work, turned much of the landscape to swamp, and made automobile travel a near impossibility.

This Packard was powered by a secret new motor, the "Twin-Six," the first-ever twelve-cylinder production engine. Apparently a little superstitious, Eisenhut said later that the twelve straight days of rain they suffered were due to the twelve-cylinder motor. He said he was glad it hadn't been a twenty-four-cylinder model.

These three were far better prepared than Post had been. Joy was a seasoned automobile explorer who had made numerous transcontinental auto trips, including some of the earliest on record. He made yearly trips of great distance to personally test the new Packards, both before and during his involvement with the highway. Like a true field engineer, Joy would stop en route and send back lengthy telegrams to company headquarters reporting on difficulties, possible remedies, and especially on the success of improvements and solutions to problems found on earlier trips in last year's model. He had had this car specially equipped to handle bad roads, mechanical breakdown, and any other possible difficulties. A much higher car than Post's European roadster, the Model 1-35 had greater ground clearance and could negotiate deeper mud and taller road obstacles. The Packard was also outfitted with electric lights, a new idea in those days that allowed Joy and his companions to travel safely at night. The car weighed five thousand pounds carrying a full camping outfit—tents, cooking equipment, and provisions.

These were true reliability tests for Packard, but they also served as an outlet for Henry Joy's wanderlust. He clearly relished these trips out across the plains and mountains. A robust, fit man, he loved the outdoors and life as an auto nomad. In jodhpurs, tight-laced knee boots, and khaki camp shirt, he looked the true explorer and felt at home camped on the empty reaches of Wyoming. Like John Wesley

Powell, like Captain Benjamin L. E. Bonneville, he labored across the open spaces reading the terrain, studying the passes, making dense notes, tasting dust and food cooked over a fire. He slept in the open in the western silence, and at night he rested to the creaks and gurgles of a cooling Packard instead of the nickering of hobbled horses heard by explorers of the past.

Stubborn and self-reliant, Joy was no stranger to hard physical work. At one point on this trip, bogged down for hours in the mud of the prairie, he worked to exhaustion before admitting defeat and sending for help to extricate the car.

As president of Packard, Joy spent most of his time wearing the stiff collars and suits of the capitalist, and tending the operation of a growing auto manufacturer in a market of growing competition. His days were spent in offices and boardrooms, dealing with sales figures, executives, and stockholders. To Joy, the mud, terrain, and distance must have seemed noble and measurable adversaries after dealing as he did with the abstractions and minutiae of business.

There were, no doubt, many others within the Packard organization who could have made these test trips and left Joy to run the company, but for many years he insisted on conducting them himself. Joy wanted Packard to make only cars that he would enjoy driving himself, and the best way to find that out was through his annual trips west. Joy, with the help of the Lincoln Highway, gave each new model or improved version of an older model a fair and sometimes extraordinary test. Packard advertised, "Ask the man who owns one." A prospective buyer could hardly find a more knowledgeable and enthusiastic owner than Henry Joy.

His desire for firsthand experience extended to his work for the Lincoln Highway Association. A. F. Bement, as new secretary for the organization, was in charge of field reporting, promotion, and guidebook preparation, but he obviously took a back seat to Joy when it came to enthusiasm. It is not clear what sort of accommodations Be-

ment and mechanic Eisenhut expected, but Joy thought little of the commercial hotels along the route, so there was no doubt that roughing it was the order for this trip.

At 9:45 that first night out from Detroit the trio neared Elkhart, Indiana, and the first red, white, and blue Lincoln Highway marker appeared out of the rain in the glow of the headlights. There the men in the Packard turned west onto the great road. Though there were numerous good hotels nearby, they made camp near Valparaiso, Indiana, in a downpour.

By morning, they were wet and the car full of water. Building a fire and cooking in such rain was out of the question even to the outdoorsman Joy, so they bailed out the car and drove into Valparaiso for ham and eggs. It was to rain for nearly two weeks.

The rains continued as they plowed their way across Illinois and into Iowa. Joy began to refer to the drive as "submarine work" and suggested that their trip be called "The Exploits of the T-6." The humor subsided a bit near Tama, Iowa, where they spent three hours in a cold rain freeing the car from one mudhole. It took them five hours to cover thirty-five miles that day. They averaged 1.9 miles to each gallon of gas.

Joy, however, was undaunted. In high spirits, he turned a shovel hour after hour, laid on his back in deep muck to dig beneath the Packard, was continually drenched by downpour, waded through knee-deep water to find firm footing for the tires, scouted lumber to pry the car from deep trouble, ate miserable food, slept along the roadside, awoke day after day in the rain, and felt "ace high."

Little moved in Iowa under the continual rain. In Boone, they were told that all the autos that had arrived in town during the last week were still there. The roads in all directions were miserable, but the road to the west was submerged. The Des Moines River had risen to cover the Lincoln Highway for a considerable distance. Game but anxious, Joy and his companions drove to the west edge of town,

*Henry Joy (left) and Austin Bement
celebrate triumph over a Nebraska mudhole
on their 1915 trip to the Panama-Pacific
International Exposition.
Courtesy of the University of Michigan.*

followed by curious residents and stranded autoists. What a story this would make: Henry Joy, the president of Packard and the president of the Lincoln Highway Association, swamped in his Packard on the Lincoln Highway!

What greeted the Packard crew did not look good. The road disappeared into the brown river for maybe a half mile. On the far side the mud road could be seen climbing the opposite hill. Marking the general course of the road through the flood was a double row of fenceposts, protruding a scant few inches above the water. The men could only guess how deep it was. While the spectators looked on expectantly, the three held a quick conference in the Packard. They estimated that the carburetor would probably remain above water, unless, of course, the submerged road was washed out.

They decided to make a run for it. "Stand by! Ready about!" shouted Joy, the submarine captain. With great care they set off through the flood, keeping the heavy Packard centered between the fenceposts, driving slowly to keep waves to a minimum. Water rose above the hubs and running boards, but the carburetor and ignition remained dry. With great relief, and amid distant cheers from the audience across the river, they left the Des Moines River behind and ascended the slippery hill headed west.

On June third, Joy cabled the Packard offices from a hotel in Gothenburg, Nebraska, where they laid up a day and a night to wait out the rain.

It started to rain just before we got to La Porte, Indiana, the first day, and it has rained every hour of the time since then, with the exception of one half day. It is raining now in floods and it is sailor's luck that we got in here for the night. Never in my life have I seen such roads. They are from twelve to twenty inches deep in gumbo. We made 37 miles on Tuesday and have been in low gear practically all the way from La Porte. Ahead of us, in North Platte, Nebraska, they say they are using small boats in the streets.

Ironically, though rain fell in torrents outside, their hotel had no running water.

Later, arriving exhausted in Cheyenne, they stayed in a hotel, the second such luxury since Detroit. Here their luck changed; the curse had run out. That next night, they camped 250 miles west of Cheyenne under clear and silent western skies to the smell of bacon frying over a sagebrush campfire. Dry, warm weather replaced the rain.

It had taken Joy, Bement, and Eisenhut eleven hard days of effort to cover the thousand miles between Chicago and Cheyenne. With good weather, they could have done it in little more than three days. They had been on the road, either at the end of a shovel or in the car, for twelve to eighteen hours each day. Now, with their struggle ended, they crossed the broad sand and rock of the desert and mountain sections with ease.

Both Emily Post and Henry Joy eventually made it to San Francisco and the Panama-Pacific Exposition. Post, disgruntled with the Lincoln Highway from the start, left it at Cheyenne. Her party found the road better beyond Illinois, but the desert Southwest was an irresistible lure with its colors and Indian cultures, far preferable to the bleak deserts of Utah and Nevada. Besides, anything was better than that miserable Lincoln Highway. She and her companions drove south through Colorado Springs into New Mexico and Arizona. As bad as the Lincoln had been, they found other sorts of difficulties driving the little-traveled wagon routes through the Southwest. Had they stayed to the north on the Lincoln Highway, their trip across the arid West would have been much easier. Their low-slung car was damaged by large rocks in the road. They feared breakdown far from help and were often lost on unmarked roads in wilting heat. Utter lack of hotels—good, bad, or indifferent—forced the New Yorkers to spend a night in the car under the stars of New Mexico.

Finally, with the car in serious need of repair and more than five

Henry Joy discovers that direction signs along the Lincoln Highway of 1915 are as informal as the road itself.
Courtesy of the University of Michigan.

hundred miles of hard road ahead to Los Angeles, they threw in the towel at Winslow, Arizona, and had the car shipped west. Meanwhile Emily resigned herself to the train and a side trip to the Grand Canyon before proceeding to the West Coast. Later, with the car repaired, they made an easy tour of the coast on California's well-maintained roads, making a brief stop in San Diego for the Panama-California Exposition and spending several days at the Panama-Pacific Exposition in San Francisco.

Henry Joy and his companions made a more triumphant entrance to San Francisco and the exposition. Their unkempt car was the center of attention as it crossed San Francisco Bay on the ferry, rumbled through the city, and entered the exposition grounds. Covered with mud and road grime, filled with dirty camping equipment, it was promptly placed on display in the Palace of Transportation, where it soon drew a crowd.

Those who gathered at the ropes to view the Packard couldn't help but see several states' worth of the Lincoln Highway caked and plastered to its wheels, running boards, and sides. The Lincoln Highway had given the Twin-Six a violent workout and it had stood up without failure. This muddy display was certainly a great testimonial for the Packard, but it was no promotion for the Lincoln Highway. Joy, who sang the praises of the car, certainly was disappointed with his coast-to-coast road. It had taken him twenty-one days to make the trip from Detroit to San Francisco. After a trip on the road two years earlier, he had predicted that by 1915, with the improvements then under way, travelers would be able to cross the entire continent in only eleven days.

The Lincoln Highway was no highway in the spring of 1915. Instead of being a completed highway to the fair, it was a mudhole that extended from Illinois to Wyoming. For the most part the route had been marked, but the little real improvement previously accomplished had been quickly swallowed up by the floods of spring. It would never be this bad again.

Some twenty-five thousand people passed through the Palace of Transportation during the two days that Joy's Packard stood there in muddy glory. Among them no doubt were many who pondered an automobile trip of their own, a trip longer than the usual Sunday drive, or something beyond the leisurely tour of nearby towns and parks. They stood and looked at this earth-colored car, resting as it did among flawless and polished new automobiles of every manufacturer. Certainly this exhibit convinced the faint of heart that the train was the better way to get from place to place. But for a growing number of people who itched for a long auto trip, it kindled the flame of adventure, it struck the spark of new experience. And even Emily Post said she'd do it again. In her 1917 book *By Motor to the Golden Gate* she wrote:

Some day we are going back. When we go again, we are going in two cars—one to help the other in case of need, and, if possible, a third car to carry a camping outfit—and camp! [Alice] and I both hate camping, so this proves the change that can come over you as you go out into the West. . . . Why difficulties seem to disappear; and why that magic land leaves you afterwards with a persistent longing to go back, I can't explain; I only know that it is true.

Few people standing there looking at the Packard knew the extent of Joy's travails, or the difficulties of Post and her companions, and it probably would have made little difference if they had. Like some holy relic, this Twin-Six was different from all those shiny cars nearby; this was a machine that took on mystical qualities because it had done something adventurous, something those gathered at the velvet rope were dreaming about, or planning.

By 1915 the automobile had become the newest vehicle for American restlessness, the newest outlet for a national wanderlust. To a people who listened with longing to trains in the night, who tossed sticks in rivers and watched them carried downstream out of sight and wondered about places downriver, to a people who relished the wind of travel on their faces and who preferred it without the concomitant cinders of railway travel, the automobile was the perfect machine.

The breeze rushing through the open windshield of an automobile was stronger than that on a boat or in a buggy, and the hiss of moving air blended smoothly with the sound of a powerful motor. It carried the perfume of motoring: the smells of rubber, oil, and gasoline, and the scents of woodland, river, prairie, and sage. It was this wind with the smell of someplace else in it that urged the traveler on, that made it clear that you were on the road for somewhere. Americans bought and drove open cars long after much quieter and cleaner enclosed cars were available. It wasn't the extra cost of an enclosed car; it was the wind.

Americans have always been a restless lot. In 1847 the South American statesman D. F. Sarmiento said, "If God were suddenly to call the world to judgement He would surprise two-thirds of the American population on the road like ants." Our national monuments have been the Conestoga wagon and the steam locomotive, the automobile and the airplane. Built of wood, built of iron, steel, glass, and rubber, American icons of travel replaced the architectural wonders of our European roots. Instead of perfecting a national style of building, we perfected the engines of travel, the machines that nibbled at the frontier until we stood at the western sea.

Some have suggested that the first Europeans landed on the coast of America and simply kept moving west. Certainly the Puritans and planters had more pressing concerns than what lay toward the sunset, but they were seasoned travelers, having already crossed the stormy Atlantic; so the seeds of restlessness, the desire of escape over the mountains, could be ignored for only a little while. Once the Appalachian barrier had been breached, the westward migration grew, fueled by free land, and was called Manifest Destiny. With little regard for displaced native cultures, Americans canalboated, turnpiked, and steamboated to the edge of the Great Plains, from where the overland wagon and the steam locomotive carried them to the sea.

Americans understood a new east-west national unity at a time of great rift between north and south when the overland mail and stage service tied the coasts. They drew ever closer when the transcontinental telegraph made communication instantaneous in 1861. The coasts were joined by iron in 1869 with the completion of the transcontinental railway—churchbells rang, bands marched, the nation celebrated for days. The line of the frontier was pierced; it would soon disappear altogether as the map of the West was filled in with towns, railroads, and new names for rivers, mountains, and lakes.

Although the free land that fueled the westward expansion was gone by 1890, the idea of the frontier lived on in the American mind. A people so used to movement could hardly settle down just because there were no more homestead claims. Americans were still restless. We were still eager for more, wanting to go on.

At the dawn of the twentieth century the automobile began to creep its way out of the garages of tinkerers, onto the streets, and into the countryside. This new machine was noisy and smelled bad, and it often quit far from home, but it didn't take very long for Americans to grasp its potential, to adopt the device and to push it to see how far it would go. Here was another outlet for wanderlust, another way to span the continent, to bind the East and West. The frontier hadn't ended; it still lay in the roadless wastes of Nevada, in the undiscovered little towns of Iowa, and in the wind that tugged at your hat when the speedometer hit thirty.

The early automobile tourists, those who took to the road before the First World War, were the pioneers in a new tradition of American travel. They sallied forth onto the roads of the country and set a pattern of transportation that is the foundation of how we travel today. They experimented with and set standards for a sort of mobility that no one had previously imagined, a mobility made possible by the automobile.

Emily Post and Henry Joy certainly demonstrated that, at times, travels could be quite difficult. But as a rule, if the weather was drier, the travails were more tame, more along the lines of occasional mudholes, ordinary breakdowns, or poor hotels. Few had as much trouble as Post and Joy. Nevertheless, Post and Joy are representative of those early motorists; like many Americans, they both had a good bit of wanderlust and a certain eagerness for adventure, and both were people of means sufficient to undertake such trips. While many of Joy's trips were bankrolled by Packard, it is clear that his station in corporate life made such adventures possible.

In the early years before World War I, and before most working-class Americans had any opportunity for any sort of vacation, it was only the leisure class who could afford to purchase and operate an automobile, buy fuel, repairs, food, and lodging en route, and afford to take sufficient time from jobs to make a cross-country trip. At a time when a blown tire was worth two weeks' wages to a working man—and several tires could go in a single day of hard, hot driving—auto touring was definitely the province of the well-to-do. They drove automobiles with imported leather upholstery, large cars with classy names like Overland, Winton, Maxwell, and Packard. These were the newest models, outfitted with the latest advances in automotive engineering, the best accessories and equipment. These were the first "all-electric cars," with lamps for every purpose, and the cars with the big, multi-cylinder engines. They cost as much as a house. In this era of open cars, travelers chose the best in motoring attire:

for men, a long leather coat, heavy gloves, and cap; for the ladies, dusters made of fine linen, calf gloves, and silk or crepe de chine veils; goggles for everybody.

In the cities along the East Coast, and especially in New York, it became fashionable to take a long automobile trip; making it clear to the Pacific would get mention in the papers. Physical barriers like the Rockies were one thing, but the likes of Emily Post could overcome even these if the New York Four Hundred looked upon the effort favorably. Though her automobile quit short of the coast, she was highly praised for her exploits, and she provided great encouragement to many others who followed.

The Lincoln Highway appeared just at a time when it was no longer enough to set sights on New England; that was old hat to the monied society of the East and had been criss-crossed and trampled since the first single-cylinder Oldsmobiles sputtered along the tow paths of the Erie Canal. Many, including Post, had toured Europe by auto, but with the war, that too was out. The auto travelers, looking about for their next adventure, the next trip to garner ink on the society page, looked clear across the continent to California. Though troubles may lay in the sand and mud of the West, the American land beckoned; "See America First" became the cry of the Lincoln Highway Association and auto touring organizations across the land.

Like Post, many travelers wrote about their exploits. Magazine editors and book publishers were hungry for tales of auto adventure on our own continent, and readers itchy to hit the road fantasized about what fun they might have. Post's book probably spurred many; at least two later travelers mentioned her book in theirs.

It's no wonder that Americans adopted the automobile with such open arms, and no wonder that so many soon clamored for better roads. On dry days it was a profoundly new way to travel. Eager travelers had a small, very mobile vehicle that could carry everything most people—even Emily Post—could want for an extended, almost unlimited trip. Here was mobility in a package that offered ultimate flexibility. It was a flying carpet. Freed from confining train schedules and routes, the motorist could make decisions about destination, time, and route on whim, and exercise them with simple ease. The automobile's untiring engine and multiple gears allowed it to jump great distances. The sight of an inviting road and the turn of a wheel opened a new path to a new adventure, and the push of a pedal or the pull of a lever brought the vehicle to a halt where no train had ever stopped.

Certainly the automobile also had less grandiose functions and supporters than long-distance touring by the well-to-do; rural doctors were among the first to buy automobiles because they made it much easier to conduct a house-call practice among scattered homesteads. Farmers flocked to the Model T Ford by the thousands. When the roads were dry, this machine made the trip to town a quick jaunt, rather than an all-day affair. But among the vanguard for "automobility" were those who set off to see the country.

For the auto traveler, the pattern of the day became a continuous unfolding panorama that moved at a pace determined by the motorist. As the car was being repacked each morning, out came the guidebooks, what primitive maps there were, and the directions scribbled the day before on the back of an envelope, given freely and in great detail by the farmer up the road.

The small wonders and dramas of common landscape and experience held the attention of autoists as they traveled at thirty-five miles per hour. A threshing bee, a brief stop at a ranch for water, a funeral procession—they were like short scenes from the theater, lifted in exact detail from some play, snippets of action quoted out of context, leaving the plot hanging. The travelers seldom stayed to see the act end; they moved on, leaving the outcome unresolved. But when Joy forded the Des Moines River before a large crowd, and when a big car with lots of equipment and luggage tied to it arrived in town, the auto travelers stepped across the floor lights onto center stage for a scene from the drama of the road. No longer did they merely observe these little acts; they became star characters.

Thornton Round was a young man just able to drive in 1914 when he traveled with his parents and family in two cars from Cleveland to California and back. On their return trip, they stopped in Wadsworth, Nevada, a prettier-than-average town not far from Reno:

. . . a crowd of people gathered around us and began to pepper us with questions. They asked where we had come from, what conditions the roads were in, whether we had seen any wild animals, the total number of miles we had traveled; they were also curious about the Winton, and they even asked what we carried under the canvas. We didn't mind the questions one bit. We found people to be very interesting in all of our brief encounters along the way. Many times strangers had been of great help to us by offering needed information on the local areas, bits of information we found invaluable, and could not have possibly known except for the kindness of the people who were in a position to know the facts of their own countryside. We had learned from experience that we could depend on the local gentry for the things we needed to know in order to plan about fifty miles ahead in our travels. When we had succeeded in covering the fifty miles or so, we would stop and make further inquiries.

At another stop, Round describes how the curious residents "very carefully examined all the pennants decorating the cars, and read the postcards we had tucked in the bottom of the inside of the windshields."

James Flagg, a New York artist who complained his way across the country some years later, also stopped in Wadsworth. His attitude was generally bad, and he wanted things to be like New York no matter where he was. He arrived after a particularly trying day in the desert: "I sat down on the running board and gazed through dust-caked eyelashes and with depressed spirits at the most miserable collection of hovels that ever had the impudence to don a resounding name like Wadsworth." It is hardly surprising that crowds didn't rush out to shake his hand and ask him questions.

Another traveler, Frank H. Clark, who traveled west from New York state in 1918, had a better attitude than Flagg: "We found acquaintances we had never dreamed of and many friendly strangers. Of the former, the most noticeable were to be found among the young Americans who, more generally than their elders, knew and could call the passing motor cars by name. 'Ah, there, you Franklin!' was a greeting from the school boys, varied in wording sometimes, but frequently heard."

Local help was often essential to finding the way and avoiding difficulty, and a day on the road was often spiced up by meeting a memorable character when asking directions or seeking help. None was more memorable than one John Thomas of Fish Springs, Utah. For many years this giant of a man lived the hermit's life in the old pony express station there, and according to some, made his daily bread by pulling stranded tourists out of his homemade mudhole smack in the middle of the Lincoln Highway. He claimed to have been an agent for the pony express and the overland stage, and was certainly old enough. He stood "six feet four inches tall and four feet wide," and always wore a decrepit, stained felt hat. Henry Joy is known to have sat at his table and shared a meal and perhaps a drink of whiskey or two with old man Thomas, and very few who ever passed his way forgot him or failed to mention him in their writings.

Auto travelers approaching the Thomas place from the east could see it for many miles across the desert. As they drew closer, within a few miles, there was a place where the Lincoln Highway forked: the right fork headed toward the shimmering oasis of Fish Springs across a low swampy area; the left fork stayed on higher ground but was obviously much longer. Here a sign warned the traveler to stick to the high road in wet weather. Most travelers, even in the wetter springtime, probably hadn't seen water for a hundred miles or better, so they took the low road. It was fine for a mile or so, but then became rough and muddy, while the landscape to either side was damp but smooth and solid looking. It was here that Thomas had supposedly diverted a small spring to flood this low spot.

A nearby sign cautioned the traveler to stay on the road, but usually the temptation was too much; the ruts looked dangerous, the salt-frosted swamp firm, and soon a heavy touring car had turned out of the road, broken through the crust, and was mired to the running boards. Looking about for salvation, motorists would then see another sign, a hundred yards ahead: "If in need of a tow, light fire." A pile of sagebrush was provided by the post.

Soon after the first puff of smoke, Thomas would appear leading two draft horses, more accustomed to pulling cars than plows, and would announce his price to pull the errant car back to the road, usually a dollar a foot. He'd silently listen to the driver argue for a few minutes, then raise the price to two dollars. Smart travelers shut up and paid the fee—the longer one argued, the higher the price.

Local legend has it that Eddie Rickenbacker once sprung the Thomas trap, and had been pulled back to the road before the old desert man mentioned his price of twenty-five dollars. Supposedly Rickenbacker blew up at the fee, so Thomas rehitched his steeds to the car and pulled it back into the muck. Mad as hell, but seeing no other option, Rickenbacker agreed to pay and Thomas pulled the car out again. When handed the money, Thomas said, "No, that'll be $50—I pulled you out twice."

He gouged the likes of Rickenbacker in his expensive car, but Thomas is said to have had a soft heart. If the stranded car was an old flivver or had a large family aboard, he was likely to refund all the money and give the driver a lecture about paying attention to road signs. In any event, the desert crust would soon smooth over after a rain shower or two and the trap would be set for the next victim.

Thomas was said to be a fine host if he took a liking to his guests and they were willing to pay the freight. Effie Gladding and her husband bought a noon meal from him at the old stage station—"fried eggs, potatoes, pickles, cheese, bread, butter, and tea, and an appetizing cup cake cut into squares." Since they were eastbound, they hadn't passed the mudhole, and which road he told them to take is unknown, but he did advise them to build a fire should they need help of any kind.

Smart travelers took the high road and missed the mudhole altogether, but many still managed to leave old man Thomas a fair bit of their money. Thornton Round's family paid fifty cents a gallon there for gasoline at a time when it went for about fourteen cents elsewhere. Another stayed the night in the cold stone hut and left before breakfast, the place having offered little more than "fighting gnats, poor food and unclean beds, innocent of linen." Those who stayed the night and curried the favor of their host heard stories—some true, many not—of the pony express, narrow escapes from Indians and desperadoes, Mark Twain, and desert lore. Round remembers Thomas saying that Brigham Young himself taught the old man to read.

Whatever the flexible fee structure for towing and accommodations, whatever the misinformation about history or the possible diversion of desert streams, John Thomas still provided a valuable service to long-distance travelers in the desert. He also added a memorable dash of color to the early road.

Wealthy easterners were generally unaccustomed to western mores and sometimes did chafe at certain western customs. Effie Gladding and her husband concluded a lengthy round-the-world trip with an auto tour from San Francisco to their home in Montclair, New Jersey, during the summer of 1914. After several weeks among the uncivilized westerners, the Gladdings arrived late one day in Carroll, Iowa, and took their supper at a cafe near the hotel.

We were interested in a party of four young people who were evidently out for a good time. The two young gentlemen, by a liberal use of twenty-five cent pieces, kept the mechanical piano pounding out music all through their meal. They were both guiltless of coats and waist-coats. We had seen all through the West men in all sorts of public assemblies, more or less formal, wearing only their shirts and trousers. So we had become somewhat accustomed to what we called the shirt-waist habit.

In the West, meeting another automobile was often a celebrated event. If the car was going the same direction, an alliance within the "motor fraternity" was often formed to help at muddy sections or just to provide moral support. If headed in the opposite direction, a great deal of valuable information could be exchanged about the condition of the roads to come, the better places to stay and eat.

In Nebraska, Bellamy Partridge met a car from California, "battered and mud-stained. We stopped and talked together reporting the conditions we had come through. We both sounded rather pessimistic, but we took pictures of each other and went doggedly on hoping for the best."

So few were the routes in the West that a westbound motorist could almost make appointments to meet and powwow with eastbound motorists. Before his trip, Partridge had corresponded with A. L. Westgard, pathfinder for the American Automobile Association—sometimes referred to as "the Daniel Boone of the Gasoline Age"—and since they would be going opposite directions at about the same time, they agreed to try to meet along the way somewhere. In the desiccated heart of Nevada, Partridge saw a cloud of dust off in the distance:

If it was a car it would be the first one we had met on the road since leaving Ogden. As it came nearer we could see that it was indeed a car. A little later one of the girls, who had either phenomenal eyesight or a good imagination, reported that she could see a red license plate. The New York numbers were red that year. Soon we could all see the red number. It was Westgard.

I pulled out of the road and stopped. He drew abreast and stopped. Before I could get out, he was over there peering into the car and shaking hands all around. "You see I'm punctual about keeping my appointments," he said.

They took photographs, discussed road information, and set a date to meet in San Diego the following year.

The end of the day again found the auto traveler in a new place, a place rich in unimagined detail, sometimes pleasant, sometimes not, a place that endured in the memory of sight, sound, and smell long after the trip had ended. It was often an unplanned place, a hotel, a camping spot on the prairie, or a ranch house.

Alice H. Ramsey was the first woman to cross the continent by auto when she and three female companions made their memorable journey in 1909 in a new Maxwell. For the most part they followed the route that later became the Lincoln Highway. One night the travelers reached Opal, Wyoming, and, intrigued by the name, decided to stay the night.

Hermine and I shared the same room always and, of necessity, the same bed. We were both good sleepers, however, so we managed all right.

I had been asleep for hours when I was aroused by Hermine's moving about considerably. As the disturbance continued, I said, "What's the mat-

ter with you?" If she had been a light sleeper she probably would have been more aware of the situation. When she answered sleepily, "I don't know. I seem to itch all over," that was all I needed to hear. I bounded up quickly, lighted the oil lamp and took one look at the bed. One look was enough! Immediately I began hurriedly to dress, and Hermine, now half awake, inquired, "What are you going to do?" I parried with, "I don't know, but I'm going to get out of here, that's for sure."

That must have sounded like a rather unsatisfactory reply, for where could we go in this section of the country at this hour of the night? The time was 2 A.M. as we made our way quietly down the stairs into the so-called office. Not a soul was around. The night clerk, at this hour, was a blackboard hanging on the wall upon which was scribbled the numbers of the unoccupied rooms! When a traveler came seeking a place to sleep (and it was likely to be a sheepherder) he consulted the board for information and proceeded to his "downy." One could only hope his eyesight was good and that he read numbers correctly! All the keys to the rooms had long ago been lost!

The women spent the rest of the night in the lobby, sitting in chairs and resting their heads on a table. Ramsey wrote, "The town's name *Opal*, was assuredly a misnomer—no jewel, to us!"

Their other experiences with accommodations were less annoying, and in fact sometimes amusing. Later on, west of Austin, Nevada, the women stayed at the ranch of Pat Walsh. After a better night's sleep than in Opal, they arose to a breakfast of lamb chops and chocolate cake.

By 1915, the automobile was on the way to becoming a reliable vehicle but still having teething fits. It was less often necessary for drivers to have to "get out and get under" as Al Jolson had sung, in order to fix the major problems that had plagued earlier motorists, upheavals that frequently stopped cars dead, and usually in the rain and far from home. Ignitions, carburetors, major engine components, transmissions, and axles had all been improved such that most cars could make a cross-continent trip with only minor repairs.

A traveler could still easily find trouble by gnashing gears when shifting or by hitting holes too fast. Metallurgy simply hadn't quite caught up with the need for light, strong parts that would take the miles of pounding. Most of the technology of the day centered around railroads, where making something stronger simply meant adding more material to beef it up; the rest applied to bicycles and buggies, where strength was a smaller concern because loads were light and parts could be correspondingly slender. A touring car easily weighed two or more tons, putting great strain on suspension assemblies and drive trains. Transmission gears, springs, and steering parts were the second things to go when the going got rough, which it often did.

What failed first were tires. While the infant rubber industry tried every imaginable concoction to improve the strength of early high-pressure tires, tourists wandered about the countryside having flats and blowouts at every opportunity. Sometimes it was the heat, sometimes sharp stones, and sometimes there was no reason at all; and in some sort of just revenge, it was very often loose horseshoe nails.

Thornton Round's family kept records of tire wear at the request of Goodyear on their 1914 trip from Cleveland to California and back. They probably weren't able to tell Goodyear much about how many miles it took to wear the tread off a tire—with two cars, a total of 7,245 miles, they had 175 flats and seventeen complete blowouts. The Rounds spent most of their evenings by the campfire repairing their collection of tubes and tires for the next day's mileage.

By necessity, most early travelers were resourceful. They carried tools and extra parts, block and tackle, food, and emergency supplies. They were often their own mechanics, reattaching things that had fallen off or changing fouled spark plugs by the roadside. Most black-smiths were far more familiar with horses and farm implements than with automobiles, so the prudent motorist took pains to know his or her machine well.

These early long-distance auto travelers made much of themselves as pioneers: free-spirited individualists who, when the condition of the roads allowed, went as the wind blew, relied on wits and a supply of spare parts and extra tire casings, blazed new trails, and took life as it came. But with the exception of A. L. Westgard, who was once stranded for sixteen days by high water on a river island, most of the adventures were really rather tame. Though bad roads were common, the trials of the likes of Post and Joy in the spring of 1915 were somewhat unusual; a month after Joy struggled to the coast, the rains had stopped and the highway, while no speedway, was at least passable.

The attraction of adventure was found not in wrestling with three states' worth of mud but in the pleasure of abandoning oneself to unfamiliar surroundings. Early auto travelers relished the closeness to the places where they traveled. From a train they saw the landscape in abstraction like a motion picture, but from an open automobile the land was made vivid by intimacy. The autoists traveled within the land rather than across it. They became part of the landscape of the continent as it unrolled before them; they happily shrank to insignificance in the canyons and deserts of the West and sat as quiet observers in a cafe watching a couple of fellows impress their girls by feeding quarters into the mechanical piano.

Certainly they were pioneers, but not in the usual sense. They didn't rely on oxen for propulsion, and few died of thirst in the desert, but before the explosion of auto touring in the 1920s they were the pathfinders, the explorers who not so much found new routes as established the patterns for a new method of travel that would become an American culture—the automobile culture.

II
An American Geography

4

New York City to Chicago: The Settled East

With a last look at the wonderful sky-line of the city, and the hum and whirl of the great throbbing metropolis, lessening in the swirl of the Hudson River, we really were started; with our faces turned to the setting sun, and the vast wonderful West before us.

— Beatrice Massey,
It Might Have Been Worse

But what of the highway itself? How did it go here; where did it cross there? What of the continent it spanned? Who traveled here before, long before the internal combustion engine?

To know the Lincoln Highway and the continent it crossed, it is necessary to look at the road in piecemeal fashion, not as a line bridging the continent in a great leap but as an end-to-end collection of different roads, paths, and routes. We must see it section by section, state by state, and in some places almost mile by mile before we can understand the history and geography of this highway.

This is the highway as place; not as a connection or as a thoroughfare between cities, but as an entity unto itself. Like a farm or a neighborhood or a steel mill, a road is a three-dimensional thing with length, breadth, and depth of history. By knowing this road as a place, and as a place within history, we can understand not only why it went where it did, but we might also understand some of its connection with history and the great attraction it has had for travelers since the very earliest days.

This highway has a story older than its establishment in 1913, a story that began long before anybody thought of naming it Lincoln or imagined driving an automobile on it. When Henry Joy connected older paths to make this highway, he also gathered historic threads of past travel on the North American continent. Many of these paths have a lore of their own, stories that detail how America learned to travel, how a nation assembled itself across three thousand miles of continent. The history of this route from east to west is a sort of chronological sampling of transportation development and western migration in America, starting with early coastwise travel along the eastern shore and progressing to the opening of the old Northwest and the push of pioneers and railroads to the West.

For much of the way across the continent, this highway followed what is loosely referred to as the central route. This broad and general path of transport is sometimes very wide and sometimes quite narrow. It is two states wide crossing New York and Pennsylvania, and as narrow as the Platte River valley in Nebraska. This central way across the continent has been the backbone of the nation's transportation systems and has seen a greater concentration of travelers than perhaps any other route.

Many older paths can be found beside or beneath the Lincoln Way; it followed all or parts of the King's Highway in New Jersey, the Lancaster Pike across Pennsylvania, the Oregon and California trails, and the route of the pony express. The Pennsylvania Railroad and the Union Pacific paralleled it for many miles. In as many places as possible the Lincoln Highway was wisely placed to take advantage of these well-established transportation corridors on its journey across the continent. In the East, the long-used routes gave the new highway the best chance of an improved roadway; in the West, the trails connected scattered towns across low passes and the least fearsome desert crossings. In regions of flat land in the Midwest, where towns were widespread and topography was of little consequence, paths of travel became spread out and diffused. Here the Lincoln Highway sought as straight a path as possible across the prairie.

At the very center of the story of older paths, machines, and journeys is the reflection of geography. Sometimes it is continent-wide, sometimes very local. This geography—be it cultural, landform, environmental, meteorological, architectural, or political—affected routes, travel patterns, and the devices and animals employed to get from one place to another. In that broad sense geography has always determined how people traveled, what they saw, and where they stopped. It determines why a road goes where it does, why oxen and mules—not horses—pulled overland wagons in the West, why towns are few in one place and common in another, and why the Lincoln Highway bypassed Colorado. Certainly Henry Joy's difficulty in Iowa and Nebraska was a reflection of a particular geography: a geography of mid-western soil type, good for farming but bad for roads, the geography of climate that gave the Midwest an abnormally wet year, and a man-made geography that blanketed Iowa with a denser network of rail lines than any other state, creating diminished interest in better roads.

And in the more ordinary sense of the word, the Lincoln Highway made a cross section of the continent's varied physical and human geography. The central route crosses classic and striking American landforms and biomes from eastern wooded mountains to rolling hills, to prairie and the Great Plains, high plateau, desert, barren salt desert, alpine mountains, and seacoast. The road passed through the centers of several major cities: New York, Philadelphia, San Francisco. And yet the tiny settlement of Tippett in eastern Nevada, population ten, sixty miles from the nearest settlement that could be called a town, was given equal billing, equal-sized type in the Lincoln Highway guides.

The Lincoln Highway, like all automobile highways, came to be in certain places for particular historical and geographical reasons. Historian George R. Stewart, in his book *U.S. 40,* described the development of automobile roads and saw roughly four types based upon the paths that preceded them. First were the automobile roads that began as foot or horse trails, sometimes of Indian origin, paths that were pushed from one settlement to the next and widened to make wagon roads. They languished in mud during the railroad-building age, and when the automobile appeared they were improved and paved, and relocated by and by. The Lincoln Highway between Elizabeth and Trenton, New Jersey, is a good representative of this sort of road.

Second, and less common, were roads that began as planned long-distance routes. Most of these roads were built during the turnpike boom in the early nineteenth century. They were planned and constructed as great thoroughfares of the day and sometimes made little reference to existing paths. After 1900, automobile-style improvements were often added directly to the existing path. Between Philadelphia and Lancaster, and in fact much of the way across Pennsylvania, the Lincoln follows such old turnpikes.

Third were roads that appeared alongside the railroad during the days when the train was the way most people traveled. Common in the Midwest and on the Lincoln Highway from Ohio through Wyoming, this kind of road followed the railroad because the rails often made the shortest distance between towns, and the towns had often been laid out along the tracks in the first place. In addition, a road along the tracks would disturb the least amount of valuable farm ground.

Lastly Stewart points to the roads and routes that came solely in the age of the automobile; they follow neither an older path nor a railroad but were a latter-day sort of road. The Lincoln Highway as first established used existing roads and wouldn't see this fourth road type until it began to be relocated in the 1920s.

With Stewart's road types in mind, we can travel this highway section by section in guidebook fashion. The thread of this trip will be the Lincoln of about 1915, though events and routes both earlier and later will play a role. Along this 1915 route, the places, views, and distances would be familiar to the likes of Henry Joy and Emily Post. Great ships are steaming through the Panama Canal and dusty touring cars are going west to the Panama-Pacific Exposition. Paving is rare, adventure high, and history lies deep along the highway.

New York City to Philadelphia—94 Miles

Via the auto ferry to Jersey City, marked city streets through Newark and Elizabeth, along much of the old King's Highway through Princeton to Trenton, where the Delaware River is crossed. The Lincoln Highway then passes through the growing villages of Oxford Valley and Langhorne to reach the heart of Philadelphia.

Times Square, in New York City, was a ceremonial place of beginning for the Lincoln Highway. There was no real need for a marked, improved highway here—the streets had long ago been paved, and anybody who owned an automobile there certainly knew how to get in and out of the city. Rather, the Lincoln Highway began at the Great White Way because it was the heart of the most important city in the nation. To begin at Times Square was to begin with style; it was a beginning of importance.

In 1915, the Lincoln Highway traveler, leaving Times Square, drove west on Forty-second street to a Hudson River ferry where the highway became a boat. For many years the Lincoln Highway both began and ended in a ferryboat ride—not until the Hudson River tunnels and the San Francisco Bay Bridge were completed would the traveler be on land for the entire trip.

From the ferry landing on the Jersey side of the Hudson, the highway followed city streets through Jersey City and Newark, to a path that was one of the oldest in the country. From Elizabeth to Trenton, some forty-five miles, the new highway followed the route of the old King's Highway that ran between New York and Philadelphia. The King's Highway had grown from an ancient Indian path, which was widened sometime after 1665 to allow the passage of wagons. In places the Lincoln Highway veered away from the old trail, but between New Brunswick and Trenton, the Indian trail, the King's Highway, and the original Lincoln Highway are thought to occupy very nearly the same path. Much later this route became part of U.S. 1, the major north-south road along the eastern seaboard, and when a multi-lane version of U.S. 1 was constructed to the east, the older road became a secondary route and today is New Jersey 27 and U.S. 206.

The idea that our highway and railway routes began as Indian paths has long been popular and not without cause. However, such origins are hard to prove. Indian paths were usually narrow, often faint trails that might change with the seasons depending on where the game was plentiful or the stream fords easier. Indians had no need of maps, and the early white travelers who followed their routes seldom kept records of sufficient detail to allow later researchers much possibility of correlating these paths with the routes of later travel. In general the trails of the Indian followed the higher ground, avoiding swamp and thicket and providing a better view of the surrounding country so the Indians could avoid ambushes. The white man, as soon as his transportation system began to develop, instead sought the riverbanks, where railroads and steamboats could call. That the Indians usually followed the most direct path is well known, but Indians may well have had different destinations than white travelers, and an Indian path may well have veered toward a ceremonial ground instead of to the falls of the river where the white man wanted to build a grist mill.

The early Lincoln Highway traveler drove a narrow road between Elizabeth and Trenton that echoed paths of almost one hundred years before, when Concord coaches rolled between Boston and Washington. Certainly the embankment had been built up, and the road was now hard-surfaced, but the road was still scattered with old taverns and inns set close to the right-of-way, and many of the bridges were still the stone arch affairs that the people of George Washington's time had crossed. The road gently rose and fell, turned this way and that like an old stage road or even a little like an old footpath. It took the motorist, comfortable on his high seat, back to the era of rough coach travel.

Earliest coach travel on the road between New York and Philadelphia was a slow, difficult enterprise. It took from five to seven days by boat and coach to travel the ninety-five miles between these growing cities. However, competition among companies quickly created a market for faster travel. About 1770 John Mercereau introduced his "Flying Machine," which cut the trip to a day and a half. The name invokes the image of a sleek, low coach, but the Flying Machine was like most public conveyances of the era: a straight-sided springless wagon fitted with backless benches and topped with a heavy cloth.

And fly it did; departing in the wee hours, dashing across creeks and meadows, dodging stumps in the muddy road, it delivered the passengers bruised, shaken, muddy, but happy. Competing companies drove hard to cut time from their own schedules, and each new record trip was advertised and boasted about to draw passengers from other lines. According to historian Seymour Dunbar, "The suffering travellers, sitting on their benches during such a record trip, and bounced about in the springless wagons like corn in a popper, clutched one another in desperation while they gasped out their admiration and delight at the privilege of participating in the memorable event."

A short time after the War of 1812, the famous Concord coach appeared on the road between New York and Philadelphia. It had evolved from the uncomfortable wagons, and it brought some semblance of comfort to the King's Highway and the other crude roads of the new nation.

Philadelphia to Pittsburgh—290 Miles

Via the route of the old Lancaster Pike and several other old turnpikes to Pittsburgh. The road passes through the towns of Lancaster, York, and Gettysburg amid rolling farmland, then climbs and descends the numerous ridges of the Alleghenies. The towns of Chambersburg, Bedford, and Greensburg are passed before the westbound traveler reaches Pittsburgh at the forks of the Ohio River.

When the first Lincoln Highway map was published, there had been two main routes from New York to Chicago and the West: the northern route through Albany and Rochester, along Lake Erie and through Cleveland; and a southern route, followed by the Lincoln Highway, through Philadelphia to Pittsburgh and Fort Wayne. Either route entailed a substantial jog north or south because the Alleghenies were a substantial blockade to direct east-west travel. Though the

Delaware Water Gap provided an apparently good doorway to the West, it proved to be a false entrance as numerous unbroken ridges lay astride the path beyond. There was no road of importance going directly west from New York, nor would there be until Interstate 80 was finished in the early 1970s.

The Erie Canal and the New York Central Railroad had years earlier established the northern route as a major pathway. This route dealt with the mountains by avoiding them; the New York Central called it the "water level route." Although it was at least a hundred miles longer than the route across Pennsylvania, it was an ancient path to the Great Lakes and was New York's traditional gateway to the West. To New Yorkers accustomed to riding the trains of the Central, it was only logical that it became the more popular auto road west.

Emily Post took this road on her trip in 1915. She felt that the hotels in Albany, Buffalo, and Cleveland would be superior to those of tiny, remote Pennsylvania towns like Gettysburg, Bedford, or the smoky industrial city of Pittsburgh.

She also feared the mountains. The south route through Pennsylvania dealt with the problem of the Alleghenies by making a frontal assault on them. Between Gettysburg and Pittsburgh it did battle with no fewer than five major folds, the highest reaching nearly three thousand feet. Many grades were both long and steep, forcing motorists to stop often to refill boiling radiators. The guidebooks said that the trip could be made by any car in good condition but warned that "too frequent or too constant application of brakes on the longer descents will soon wear out ordinary linings, and often render them useless before reaching the next level." Certainly care was called for.

Most likely it was Henry Joy who decided on the Pennsylvania route for the Lincoln Highway. Though it passed through far fewer cities of any size—something that might have seemed a disadvantage to Fisher—it was shorter, and Joy certainly wasn't afraid of mountains. He also realized that others had traveled this route for many years and that there was good potential for an excellent road.

In Philadelphia the Lincoln Highway meets up with and becomes coincident with present-day U.S. 30. It is often thought that 30 and the Lincoln are the same everywhere, but in fact they share identities for only two-thirds of the distance across the country. U.S. 30 begins in Atlantic City, New Jersey, and ends at Astoria, Oregon. It generally shares the route and history of the Lincoln Highway for some two thousand miles between Philadelphia and the little town of Granger, Wyoming, where 30 turns northwest to enter Idaho.

Between Philadelphia and Pittsburgh, the 1915 Lincoln Highway traveler drove the general routes of several old turnpikes: the Philadelphia-Pittsburgh Pike, the Chambersburg and Bedford, the York and Gettysburg, the Lancaster and Susquehanna Turnpike, and for many miles the exact path of the famous Lancaster Pike.

Although it wasn't the first turnpike in America, the Lancaster Pike must be recorded in history as the first sizable attempt at major road improvement. Its completion in 1795 marked the real beginning of the turnpike era that lasted until the railroads began to draw traffic away in the 1840s. The turnpike extended some sixty-six miles west from Philadelphia to Lancaster. It was built to bring farm products from the fertile soil around Lancaster to the growing city of Philadelphia and to provide an important connection to the West. Though it ended short of the Alleghenies, the Lancaster Pike was an opening to the West, drawing people to rich and foreign-sounding places like Ohio and Kentucky. It pointed the way for many who improved existing trails through the mountains to Pittsburgh and the headwaters of the Ohio.

The Lancaster Pike was built as a true superhighway of the day, and it would compare favorably with roads of today. The right-of-way was fifty feet wide, the road itself twenty-four feet wide and built of a crushed-limestone base with a smooth rolled-gravel surface. At no point did the grade exceed 7 percent. This broad boulevard impressed travelers, teamsters, and stage drivers alike at a time when the average road was improved only to the degree that the larger stumps had been removed and that in especially muddy places one might find a few branches thrown in to help stabilize the morass.

By 1820 it was a grand success as colorful conveyances of all sorts sped over the much-touted road. Nine tollgates were erected along its length, and few minded the charge for passage on such a good and fast road. The well-to-do traveler rode high and comfortably in a stagecoach—not like Mercereau's Flying Machine but an enclosed vehicle, a Concord coach—supported on springs and containing cushioned seats and curtained windows and propelled by fast horses. Freight was moved by giant Conestoga wagons, with upward-curving sides to prevent goods from sliding about, hauling as much as six tons each. A Conestoga was usually pulled by six powerful horses, and its undercarriage was painted bright blue and the main body red. Arching over all was a gleaming white fabric top. A varied cavalcade of horsemen, pedestrians, carts, drovers, and herds of domestic stock mixed with the coaches and the Conestogas—people who were headed west to Pittsburgh and down the Ohio River to new lands in the interior.

The Lancaster Pike bears a strong resemblance to a turnpike built some 140 years later—the Pennsylvania Turnpike. Both were built as gateways to the West and both far exceeded the usual standards for road construction of their day. In 1940, at a time when common highway standards called for little more than a narrow right-of-way, sharp curves and grades, and two lanes, each little wider than the average truck, the first section of the Pennsylvania Turnpike was completed. This superhighway was built with a two-hundred-foot right-of-way, four wide lanes, broad curves, and grades no steeper than 3 percent. From opening day on, it proved very successful and quickly attracted heavy traffic. Like the Conestoga teamsters before them, truck drivers found they could nearly double their payloads when they used the gentle grades of the new road. Although outmoded today, the Penn-

sylvania Turnpike set a new standard for highway construction in America and became the model for today's interstate highway system.

One hundred and eighty years before the Pennsylvania Turnpike, the route across the Alleghenies to the forks of the Ohio where Pittsburgh now stands was a prehistoric Indian path known by the whites as the Allegheny Path. It provided a steep but comparatively short route to this strategic place. In 1758, during the French and Indian War, the general course of this path was followed and widened by General John Forbes. He struggled west with some 7,500 troops and seized Fort Duquesne, at the forks of the Ohio and the site of present-day Pittsburgh. What Forbes hacked out of the forest could hardly be called a road, but after the defeat of the French the new land to the west beckoned, and Forbes Road became the backbone of this important route.

By 1818 stone-surfaced turnpikes had been built all the way to Pittsburgh, drawing a swelling flow of settlers across the mountains to fill the Ohio Valley. This end-to-end connection of turnpikes, known as the Pennsylvania Road, became one of the four major routes to the expanding West.

The 1915 Lincoln Highway traveler between Philadelphia and Pittsburgh angled around the same curves and raised the same dust as did the Conestogas of 1800. The turnpikes had fallen into disuse after the coming of the railroad; toll receipts dropped and the roads deteriorated as maintenance languished. They had reverted to little more than old wagon roads, still scattered with remnants of the grand days on the pikes: tollhouses, stone mileage markers, and old inns, most having seen better days. Their decay must have seemed picturesque and quaint to the auto traveler. Some had been modernized, but travelers felt that such improvements tarnished their charm. Many of the tollgates were still there, but the auto driver paid for a much poorer road than did travelers of a century before.

By 1916, however, the road between Philadelphia and Pittsburgh was being worked on again. Popularity of the highway had increased to the point that repair and improvement were necessary. All of the route across Pennsylvania was being paved or bettered in some fashion. Crews were beginning to unbend some of the mountain kinks and to ease the grades.

There were tolls on the eastern Pennsylvania section of the Lincoln until 1918—the only such place along the 3,300 miles of road. Under considerable pressure from the Lincoln Highway Association, the Commonwealth of Pennsylvania finally bought the road from the nearly moribund turnpike companies and took down the tollgates, allowing free passage for the first time in over a hundred years.

Pittsburgh to Fort Wayne—314 Miles

Via East Liverpool, Canton, Mansfield, Upper Sandusky, and Van Wert. The Lincoln Highway emerges from the high hills and valleys of the East into farmland that gradually flattens to a level plain in western Ohio. This is the gateway to the Midwest, where dense forest, mountainous land, and crooked roads give way to farms, industrial cities, and roads that follow section lines.

Once beyond Pittsburgh, once beyond the effects of the folded Alleghenies, the Lincoln Highway was for a time free of the constraints of rugged topography and of old, long-ago blazed trails and narrow corridors of travel. West from Pittsburgh to Canton, Ohio, the Lincoln still wound among the hills and valleys of the Ohio Valley for a while, but it soon emerged onto the level plain and farmland of Ohio, pointing toward Indiana and the West.

Perhaps somewhere west of Massillon, or past Crestline, where the terrain flattened out, the 1915 motorist on the way to the Panama-Pacific Exposition might open the throttle, might let her out a bit for the first time on some straight stretch of road.

Henry Joy, in selecting the route for the Lincoln Highway, had consistently sought the most direct path. As much as was possible, he wanted it straight and short in order to best speed the traveler to his or her destination. In Ohio the potential for a nearly straight road was best, but even here the route to Fort Wayne couldn't be a straight line; the roads with evidence of any improvement ran from Canton straight west to Mansfield, but then dipped south through Marion, Kenton, and Lima, and then to Fort Wayne.

Though this slightly longer route initially was selected as the Lincoln Highway, Joy soon changed the route to follow a straight line between Mansfield and Delphos. He could not abide the extra half dozen or so miles of this southern route, but more importantly he felt that the citizens of Marion and Lima were apathetic about the highway and hadn't offered sufficient financial or civic support. This rerouting inflamed the citizens of the towns now left off the route, and to confuse matters further, the Lincoln was soon moved again to a third route that turned out to be longer than both of the others. It was moved back to the northern and straightest route a few years later.

The exact—or even approximate—route was in hot debate for several years, and by 1921 the citizens of Marion had named the southern route after the president and their hometown boy: Warren G. Harding. Because of the popular name and the strident efforts of the snubbed townspeople, the Harding Highway got the state improvement funds and the Lincoln languished in mud. Even by 1926 the Lincoln still was unimproved dirt between Upper Sandusky and Delphos.

The division of routes continued for many years after the Lincoln Highway in Ohio became U.S. 30. Even into the 1970s Joy's straight route was known as 30N, and the route through Marion and Kenton was 30S.

These roads between Pittsburgh and Fort Wayne were assembled from the parts of many roads long used by travelers between Ohio towns and cities. Some may well once have been Indian paths, but this assemblage of loose roads lacked the historic or topographic spirit of the Lincoln Highway route to the east or west. These were roads that crossed gently rolling farmland, occasionally cut by shallow wooded river valleys, gradually growing more flat and straight farther west. Between Upper Sandusky and Cairo, Joy's northernmost route runs true as a surveyor's line for better than fifty miles. As the WPA guide to Ohio put it, "West of Mansfield the terrain is flat; and mile after mile of bright sweeps of grain and pasture land pass in lulling uneventfulness."

Fort Wayne to Joliet—185 Miles

Via Ligonier, Elkhart, South Bend, and Valparaiso. The Lincoln Highway takes the motorist across northern Indiana in a broad loop that roughly follows a handful of ancient Indian and trapper trails. Farm country gradually gives way to industry and commerce as the Chicago area is approached.

A straight-line path also beckoned between Fort Wayne and the Chicago area. The topography presented no particular challenge, but the lack of a suitable road did. It was possible to travel the straight-line route from Fort Wayne to Chicago, some 145 miles, but only by way of the Pennsylvania Railroad. From Fort Wayne, the railroad struck a west-northwest path that took it straight to Gary, where it curved around the lobe of Lake Michigan and entered Chicago. The Lincoln Highway took a circuitous path, looping northwest from Fort Wayne through Elkhart, South Bend, and La Porte, then southwest through Valparaiso and around the bottom of the lake, adding up to some 195 miles between the same cities.

The straight-line route along the Pennsylvania through Warsaw and Plymouth was a very suitable one for both railroads and highways, but in 1915 only a poor dirt road existed. The Lincoln Highway had to abandon a straight path in favor of an improved path, in this case a road which, like the path across the Alleghenies, had been a route of common usage for countless years.

The loop to the north traveled the approximate paths of bits of the early Chicago to Fort Wayne Trace, the Chicago to Detroit Trace, and part of an early Indian trail, the Sauk Trail. It long had been a route of travel, following gentle terrain, connecting easy fords, early settlements, and existing trails. It wasn't until the late 1920s that a paved road was built parallel to the Pennsylvania Railroad, eliminating this long but historic detour. The straight route was dubbed U.S. 30 from its beginning, and the Lincoln Highway moved to the new alignment along the railroad with the completion of pavement. To follow the early route, today's traveler follows U.S. 33 west from Fort Wayne to South Bend, then U.S. 20 and Indiana 2 to Valparaiso, where Indiana 2 rejoins U.S. 30.

The Lincoln Highway never actually entered Chicago. The committee that selected the route knew even in 1913 that taking the long-distance motor traveler through Chicago would result in considerable delay. Instead, the association routed the road around to the south and west of the city, staying far outside the built-up areas of congestion. For travelers who wanted to see the Windy City, routes from the south and west were marked to the center of town, but the main route of the Lincoln Highway stayed far afield. What the association created may have been the first of something that would become very familiar to auto travelers fifty years later: the urban bypass.

The loop stayed about thirty miles from downtown, but even at that distance the gravitational pull of Chicago was apparent. Between La Porte, Indiana, and Geneva, Illinois, some 115 miles, the Lincoln Highway crossed numerous rail lines leading to Chicago. Each set of tracks was crossed at right angles as the Lincoln Highway traveler went around, and the Chicago-bound rail lines went toward, the city. Alice Ramsey, who traveled in 1909, wrote:

Raymond and Anna Sipe watch traffic pass on U.S. 30 from the doorway of their SOHIO *station in Oceola, Ohio.*

Tourist cabins, Wyandot County, Ohio.

*We were given two one room shacks, each
with a double bed, a wooden stand with a
cracked pitcher and bowl and a mirror that
had a close resemblance to a babbling
brook—in a brook one's reflection might
have been more agreeable. We had supper
in an assembly with a group of fellow tourists. The scene was very rustic, a quiet
glenn—tame deer that ate from your hand
and dogs, barking and playful—cats sleepy
and owls hooting far into the night.*
—Caroline Rittenberg, 1925
    Motor West

Williamstown, Ohio.

Hubcap sculpture, Allen County, Ohio.

Unless you really love to motor, take the Overland Limited. If you want to see your country, to get a little of the self-centered, self-satisfied Eastern hide rubbed off, to absorb a little of the fifty-seven (thousand) varieties of people and customs, and the alert, open-hearted, big atmosphere of the West, then try a motor trip. You will get tired, and your bones will cry aloud for a rest cure; but I promise you one thing—you will never be bored! No two days were the same, no two views were similar, no two cups of coffee tasted alike. In time—in some time to come—the Lincoln Highway will be a real transcontinental boulevard. But don't wish this trip on your grandchildren! The average motorist goes over five thousand miles each season, puttering around his immediate locality. Don't make a "mental hazard" of the distance. My advice to timid motorists is, "Go."
—Beatrice L. Massey, 1919
It Might Have Been Worse

Churubusco, Indiana.

IF YOU
DON'T KNOW
WHOSE SIGNS
THESE ARE
YOU CAN'T HAVE
DRIVEN VERY FAR
BURMA-SHAVE
—Burma-Shave signs, 1942

Near Ligonier, Indiana.

Dunlap, Indiana.

St. Joseph County, Indiana.

Joliet, Illinois. U.S. 30 begins in Atlantic City, New Jersey, and ends at Astoria, Oregon. Until recently, U.S. 6 was also a transcontinental highway; it began at the tip of Cape Cod, Massachusetts, and ended at the Pacific Ocean in Long Beach, California. But today, U.S. 6 ends in Bishop, California, 270 miles short of the coast.

Kane County, Illinois.

The first day out of Chicago was bright and windy. The buffalo peas were blooming in pink and purple masses, and the meadow larks were singing straight at the sun. Hatless and sunburned, and feeling reck- lessly gay and relaxed, we skimmed along in our well-behaved Henrietta.
—Beth O'Shea, 1920

Rochelle, Illinois.

Twenty four hours in a town like this and we feel as though we knew it and the people intimately. In many ways it suggests a toy-land town. Its streets are so straight and evenly laid, its houses so white and shining, its gardens so green, its shops so freshly painted, its displays in the windows so new, and its people so friendly.
—Emily Post, 1915

Colonial bread sign near Dixon, Illinois.

Smith Brothers General Store, Clinton, Iowa.

Every town through the Middle West seems to have a little grill of brick-paved streets; a splendid post-office building of stone or brick or marble; a court-house, but of an older period generally; and one or two moving picture houses; two or three important-looking dry-goods stores, and some sort of hotel, and in it a lot of drummers in tilted-back chairs exhibiting the soles of their shoes to the street.
—Emily Post, 1915

Abandoned Lincoln Highway, Cedar County, Iowa.

U.S. 30, Cedar Rapids bypass.

Tama County, Iowa.

The talk was all of roads and routes and distances and interesting places thousands of miles apart, with a spice of harmless exaggeration about the performance of this car or that, its wonderful tire mileage and surpassing power and reliability. This was as natural as the elastic statements of a trout fisherman.
—Ralph Paine, 1913

Back in grade school we had learned that Chicago was a large railroad center. We believed it, all right, but that day we had a practical application of the fact which would make the knowledge stick in our minds forever. For miles and miles we crossed tracks—singly, in pairs, and in groups; back and forth we wound around and over them until we almost got dizzy and wondered if there would ever be an end. Bump! Bump! up and down, again and again. Springs were sturdy or they never would have endured such jouncing.

The bright steel of the railroads—the Baltimore and Ohio, the Erie, the Illinois Central, the Santa Fe, the Burlington, the Chicago and North Western, and many others—took aim at the heart of Chicago; the rails pointed to the city center just as a compass needle points to the north pole, marking the position of the heart of this city of transportation.

Chicago lay at a strategic point of intermission between the first and second acts of a three-act play of American travel experience. For travelers from the East Coast headed to the fair in San Francisco, this first third of the coast-to-coast trip was a shakedown cruise, an easy trial run to test equipment and skills. From here the going would be tougher, the towns farther apart, the countryside much less like home. By the time Chicago was reached, the characters all had been introduced, the moving scenes and sets established, and the ascension toward climax had begun. Here, the westbound Lincoln Highway traveler usually stopped for a day or two. She or he was nearly a third of the way across the continent, and nearing the West the plot soon would grow thicker.

5

Chicago to Salt Lake City: Where the West Begins

Any old kind of an automobile can get from New York to Chicago, and tourists have nice hotels to stop at and everything is lovely. But west of Chicago, the real trip commences. We were motorists as far west as Chicago. Then we became pioneers.
— Victor Eubank,
"Log of an Auto Prairie Schooner," *Sunset,* February 1912

At Geneva, Illinois, where the Lincoln Highway turned west again after rounding Chicago, a change occurred in most early Lincoln Highway travelers. From this point on they began to see with a different eye, to have a different outlook. Instead of seeing only the road in front, or the farms and houses nearby, instead of commenting on the architecture of the towns they passed, or the hotels they stayed in, they began to see the landscape. They lifted their eyes and began to look into the distance, away toward the horizon, anticipating that great change in the land that was soon to come. Most knew that it wouldn't come for a while—across a state or two perhaps—but once they left Chicago, they were looking for the beginning of the West.

The East and the Middle West as far as Chicago seemed a lot like home to most eastern auto travelers. The landscape, architecture, and people didn't differ greatly from those at home, so the sense of newness and novelty in travel was less intense. For many, then, the trip didn't really begin until they set out westbound from Chicago, across the plains of Illinois.

After the easy warm-up on the comparatively good roads of the East, travelers felt experienced and ready for the coming challenges. Like Emily Post, many had developed routines for keeping dust out of their clothes or had learned to judge the reliability of verbal directions so freely given everywhere. Those who camped had perfected the self-confident air necessary to set up a tent smoothly while other campers were watching, and could usually make a competent meal over a campfire or camp stove.

In the early days of the Lincoln Highway many westbound travelers made a stop of a few days in Chicago. Like Post, they shipped home items they had found to be useless, and they purchased food and utensils, extra tires and tubes, tools, and western guidebooks. They had their automobiles serviced and laid in a supply of parts that seemed likely to break or had broken once or twice already. Motorists and mechanics looked at such items as springs, ignition parts, belts, and miscellaneous fittings because they often gave out or vibrated loose and disappeared during hard travel. They put up for a couple of days in Chicago's better hotels, ate well, and slept late. They were justifiably proud of making it to Chicago but knew that the remaining two-thirds of the trip was by far the hardest part.

Most were a little nervous about the road ahead. They had been traveling among the frequent cities and towns of the East with good hotels and dining rooms and knew that they would have to be less particular in the miles to come. Everybody knew about the miles of mud in the Midwest, the sand of the desert, the silent mountains, and the long, fearful reaches of empty road where towns and possible repairs were very far apart. Emily Post's fears came to reality quite quickly after she left Chicago: the mud began only thirty-six miles from downtown. It's no wonder she was discouraged and wanted to put the car on a westbound train.

The edginess of uncertainty about the trip ahead heightened anticipation and awareness of all matters related to the adventure. Travelers turning west at Geneva seemed to lean forward in their seats in hopes of missing nothing. The senses seemed finely honed to every noise, sight, or smell. The countryside of Illinois really wasn't much different than that of Indiana, but by having Chicago behind them, everything had changed.

Joliet to Omaha—530 Miles

Via Aurora and Geneva, where the Lincoln Highway turns west through Rochelle, Dixon, and Morrison. At Clinton, the Mississippi River is crossed. The cities of Cedar Rapids, Marshalltown, Ames, Denison, and Council Bluffs see the traveler across Iowa to the Missouri River. This is the heart of the farmland where the traveler crosses wide rolling prairie turned to cultivation, punctuated by wooded river valleys and agricultural cities and towns.

For most of the way across Illinois and Iowa, the traveler of 1915 drove through gently rolling, pastoral countryside dotted with farmsteads, cattle, fields, and small towns with water towers boldly announcing town names. Broad prairie uplands brought long views uninterrupted by trees. The road crossed numerous wooded stream valleys, large and small. When the weather and roads were dry, travelers noted the passing landscape with pleasure and waxed idyllic describing what their hearts felt. Said Beth O'Shea: "The first day out of Chicago was bright and windy. The buffalo peas were blooming in pink and purple masses, and the meadow larks were singing straight at the sun. Hatless and sunburned, and feeling recklessly gay and relaxed, we skimmed along in our well-behaved Henrietta."

Dallas Lore Sharp, like many others, fell in love with the landscape of Iowa:

Near and far to the verge of the sunset the impressive homesteads stood in order as we bowled along, not casually disposed, but regularly, spatially set in the ranging landscape as if obeying some natural law. So encircling was the human scene, so distantly repeated, so balancing the picture and blending with the theme of Nature, as to make one think these houses had grown here, planted when the rich, mellow prairie was spread out for the plough.

Nowhere in the world have I felt a more perfect harmony between earth and man than among the farms of Iowa, nor more comfortable space and spiritual freedom between man and man.

Such mammoth barns! And such broods of lesser buildings roundabout! Close by stands the house beneath a group of planted trees, often barricaded from the sweeping winds by outer walls of trees. Overtopping the trees and the roofs rise the windmill and the twin silos, adding a touch of Norman times, towers and keeps to guard the wide, flanking fields and the rich booty of the yards below: cribs of eared corn, stacks of hay and weathered chaff, swine and burly cattle, horses, mules, sheep, chickens, and machinery to outfit a feudal lord.

But when the roads were wet, as they were for Emily Post, the scenery and euphoria vanished and the solid security of the railroad beckoned. Had Post given up her trip in the mud near Rochelle, she would have had an easy time putting her car on a westbound freight. The highway leaving the Chicago area closely followed the rails of the Chicago and North Western Railway through De Kalb, Rochelle, Dixon, and Morrison to the Mississippi and beyond. The Lincoln Highway followed this railroad and its connecting line, the Union Pacific, nearly 1,500 miles to Echo, Utah, a few miles short of Salt Lake City.

Laying out an auto route parallel to a major railroad made good sense in the early days of the century, especially in the West. Towns could be found at closer intervals along the rail lines, and travelers then stood a better chance of finding accommodations, supplies, and repairs; and should all else fail, it was easier to put the car on the train for home or another destination. The railroads invariably sought the gentlest grades and avoided topographic difficulty, factors also important to auto drivers.

Surprisingly, Henry Joy had found Illinois and Iowa to be the most difficult states in which to find the best route. It was clear to him and the others involved in route selection that the Lincoln Highway should pass through Omaha as the best gateway to the Platte River route and the West. What was most difficult was deciding what route to use to get from Chicago to Omaha. There were six main-line railroads between those cities, and several had a parallel road of some sort or another, but because of the well-distributed towns and the relatively flat terrain, Joy wasn't much concerned with proximity to the iron rails. Seeking the best route between these cities, Joy had traveled back and forth across Iowa at least ten times in five years. He felt that there were as many as fifty possible routes across the state. They were all about equal in length and in the way of topographic considerations, but they also were uniformly unimproved and tended to zigzag all over the map following a haphazard arrangement of section-line roads. With maddening randomness, all of the routes would head straight west for a while, then jog north or south to avoid a stream, to pass through some town, or to stay close to the undulating railway. Consequently, all possible routes were a good deal longer and less direct than any route that might have approximated a straight line.

The route Joy selected between Chicago and Omaha had little significance as a through road prior to the coming of the automobile. It closely paralleled the Chicago and North Western Railway all the way, and was as direct and well maintained as any, but it was by no means a standout among poor options. What pulled Joy to this route was simply that it had grown to be the route of common usage. Long before any Lincoln Highway markers went up, this path had been used by people traveling cross-country in automobiles. In 1903 H. Nelson Jackson, Sewell Croker, and Bud the bulldog came this way on the first auto crossing of the continent; five years later the famed Great Race—the New York to Paris Race of 1908—came this way. By the time Henry Joy came looking for his Lincoln Highway route, the Iowa portion had been dubbed the Iowa Official Trans-Continental Route by forward-thinking road men in towns across the state.

Omaha to Cheyenne—540 Miles

Via the valley of the Platte River, gateway to the West, through Columbus, Kearney, and North Platte. At Big Springs, the road leaves the South Platte and soon follows Lodgepole Creek through Sidney, Nebraska, to Pine Bluffs and Cheyenne, Wyoming, at the foot of the Rockies.

It began when the voice of the meadowlark changed from soprano to contralto, when the hot dog stands gave way to barbecue sandwich signs, when the garage man greeted you with a warm smile, shouting things like "Howdy!" and "You bet!" And according to Frederic Van de Water,

. . . when you see your first jack rabbit, spread out like a rumpled little fur rug upon the road; when dickcissels chant their derisive song from the telephone wires; when the black and white lark buntings become plentiful as sparrows; when half the cars you pass bear California licenses; when the auto tramp becomes a problem and the voice of the road liar is heard in the land, then you may surely know you have left the East behind you.

But for Bellamy Partridge, it began at a specific place: halfway across the Missouri River bridge connecting Council Bluffs and Omaha. A white line painted on the bridge indicated the Iowa-Nebraska border, and for Partridge, it marked the "place where the West begins." Others sensed the change back when they crossed the Mississippi, taking Iowa out of the Midwest farm belt and placing her among the cowboys. Dallas Lore Sharp, who traveled in 1927, felt it there: "At the first touch of Iowa everything looked suddenly different. The car began at once to act differently—Westernly, if you will. We began to feel Westernly, and to breathe big, Western breaths, and to gaze in a long Western way." Some were more patient and left the West to begin at the first glimpse of the western mountains; another made generalizations about western cities when describing Van Wert, Ohio.

Wherever the West began for the auto traveler, it was seldom unnoticed, because it was an important change for the westbound tourist. Sometimes precise to the width of a river, sometimes observed in the crossing of an entire state, beyond this point, farms became ranches, hired men became cowboys, and any hill might just be a butte.

Those who sensed the beginnings of the West in Nebraska had

good reason. Even the unobservant could hardly miss the gradual but profound change in the land. Frederic Van de Water saw the state as a zone of transition:

Nebraska carries the motor tourist out of the last of the East and into the beginning of the authentic West. Its corrugated roads possess no apparent upward incline, but they lift the traveler imperceptibly from a one thousand to a five thousand foot altitude in the length of the state. Nebraska is the green of earnest cultivation at its Missouri Valley end and fades out into tan and dun—the feline hue of savage, untamed things—at its other. From carefully tended farms it runs at last into the open prairie, a foretaste of the sun-soaked, wind-brightened immensity of brown empty land that is Wyoming.

Few westbound travelers could help but notice how the land began to stretch crossing Nebraska. Slowly, mile by mile, the landscape expanded, everything grew farther apart—the towns, the farms, the trees, even the clumps of grass. It was as if the ground had been pulled in all directions like an elastic fabric, putting greater space between everything. The horizon receded as the land rose and grew more flat, but became ever more sharp as the air cleared with increasing dryness.

Travelers planned on lengthening their stride as they crossed the state. Emily Post had been eager to reach the "continental speedway" of the Lincoln crossing Nebraska and looked forward to a fast dash to the mountains. She had been told that there were few stopping places but wrote: "What matter is that to you? You are not thinking of stopping at all. North Platte, perhaps, yes. Three hundred and thirty miles in a day is just a nice little fast road run." It had not yet rained sufficiently to cause the mud that Joy encountered a short time later, but instead of mud, Post found that they were limited by law to a speed of eight miles per hour in towns and a maximum of twenty in the country. "A straight, wide road; not even a shack in sight—and a speed limit of twenty miles an hour," she complained.

Nebraska also marked the beginning of the West because of the radical change in travel patterns that occurred at her eastern border. No longer were travelers in the wheel tracks of the farm wagon and the plow horse; they had entered the land of romantic themes in western travel—the overland stage, the forty-niners, the pony express, the Union Pacific Railroad.

For all but about twenty-five miles just west of Omaha, the Lincoln Highway followed the Platte River drainage system upstream to the Wyoming line: the Platte River to North Platte, the South Platte River to near Big Springs, then Lodgepole Creek to just a few miles short of Cheyenne, Wyoming. Here the Lincoln Highway was an insignificant, Johnny-come-lately addition to a broad and famous corridor of travel, for this Platte River route had long been the most important gateway to the West.

It was a very natural path for California- and Oregon-bound wagons, for the short-lived but famous pony express, for the transcontinental telegraph and the first transcontinental railroad, for an automobile highway, and much later, for lighted airways of early airmail and air passenger travel. It would eventually hold the twin ribbons of Interstate 80, and jets bound for either coast would pass seven miles above.

The valley had long been an easy route for travelers going west, leading them across the Great Plains and through the easiest crossing of the Rocky Mountain barrier, the broad South Pass crossing in southern Wyoming. Here the alpine Rockies of Montana and Colorado fell away to become a broad gateway, an arid plateau marked by buttes and open land, instead of impassable snow-clad mountain crags. The Indians had long made use of the Platte River route because it provided good grass and water and a better chance of finding firewood

than the high prairies to the north and south. They also knew that the shallow valley offered some protection from the frequent and severe storms of the plains.

The fur traders had established a path here by 1813, and the faint trail saw only the occasional red man and trapper until 1840, when some thirteen emigrants first crossed to Oregon and began the greatest overland migration in United States history. From "jumping-off places" like Council Bluffs, Iowa, and St. Joseph and Independence, Missouri, emigrants funneled toward the Platte valley, converging near Fort Kearney. From there to Ogallala, a distance of some 150 miles, the Platte River route saw the main flow of thousands of westbound settlers, bound for the rich farmland of Oregon, the goldfields of California, and the promised land of Utah.

They began as a trickle, a few adventuresome families, making the journey in hopes of better land in the West, and soon swelled to a torrent as word spread back east of fruitful lands in Oregon and gold in the mountains of California. In the peak year of 1852 an estimated seventy thousand people emigrated west using this Platte River route. The popular image of scattered wagon trains amid vast western wastes does not accord with reality. Between 1849 and 1853 most emigrants longed for relief from crowded trails and campsites, clouds of dust, overgrazed grasses, and long waits at river crossings. Like a great flood, they spread out as they traveled up the Platte valley. The trail was a broad trace of ruts and intertwined wheel tracks as overlanders picked their own paths and in some areas fanned out clear across the valley. On one occasion in 1852, so many wagons traveled west from St. Joseph, Missouri, that they were seen moving twelve abreast across the prairie of Kansas. Disbelieving Indians talked of moving east; they imagined that all the whites were going over the mountains and that the land east of the Mississippi would be deserted.

Before Indian hostilities and the Civil War slowed the westward flow in the early 1860s, nearly 300,000 settlers, miners, adventurers, and badmen pushed west up the Platte valley.

But just as the struggle between North and South was reducing the flow of emigrants, it was also bringing a new sound to the valley of the Platte and the trail beyond. As the creak of wagons and the lowing of oxen diminished, they were replaced by the thunder of fast horses and the rumble of overland stages, the gallop of the fleet ponies of the pony express, and soon after, the alien hum of the wires of the telegraph, vibrating in the prairie wind.

The pony express, the telegraph, and the transcontinental railroad all sprung from a need during the Civil War to bring the new state of California closer to the contiguous states far to the east. At a time when the Union needed to bind the states tightly into a united whole, California stood alone, nearly two thousand miles to the west.

At the outbreak of the Civil War, communication and transport between the eastern states and California was conducted via ship around Cape Horn, or by ship and land across the Isthmus of Panama, or entirely by land using the Butterfield stages that ran west from St. Louis. The ocean route to Panama carried the heaviest traffic in goods, mail, and passengers. It was a long trip involving a difficult land journey through the jungles of Panama. Depending on winds and storms, the trip took three to four weeks. But the land journey was no better: the Butterfield Overland Mail stages took twenty-five days. The coaches left St. Louis and angled far south and west through Texas, entered California at the southeast corner, and from there went northwest to San Francisco. This route avoided the winter snows of the South Pass route but added some 350 miles and several days to the journey.

Even before the Civil War began, most people agreed that a railroad to the Pacific was the best way to bring California in closer to the Union, but during the war the men and resources necessary to build it across nearly two thousand miles of unsettled country could not be spared. A transcontinental telegraph line was also being planned, and although far less costly than a railroad, it would still take time to link the East with California.

So began the pony express. It was established in 1860 as a temporary measure to unite the country by means of rapid communication in wartime. Though it has long formed a cornerstone of western lore, the pony express lasted less than nineteen months—time enough to span the West with copper wire.

The mail was loaded into *mochilas,* or saddlebags, in St. Joseph, Missouri, and tossed onto fast horses for the nearly two thousand mile journey to Sacramento. The riders were small, brave men who valued adventure over longevity. The route followed the Oregon-California Trail through South Pass to Fort Bridger in southwest Wyoming, thence to Salt Lake City. From there it struck west and somewhat south around the southern tip of the Great Salt Lake Desert and across central Nevada to Carson City, and over the Sierras to Sacramento. The route was direct and avoided the worst terrain wherever it could. Except for crossing the high Sierras, which was unavoidable, the route passed through generally open country. It was no wonder, then, some fifty years later, that Henry Joy would select much of this same route for the Lincoln Highway.

The pony express brought California two weeks closer to the rest of the United States. The fast riders carried letters between St. Joseph and Sacramento in just ten days. Telegrams were dropped at Carson City and transmitted by wire to Sacramento and San Francisco, saving additional time.

Spread all along the route at ten-to-fifteen-mile intervals were stations where horses and riders were changed. The steeds worked at full gallop; fifteen miles would exhaust even the best horses, but the riders continued for an average of seventy-five miles. Speed and reliability were all that counted. The horses were carefully selected for their fleetness and endurance; the riders were small men who traveled lightly armed. Solitary riders at great distance from civilization were at considerable risk from Indian attack, but the strategy called for flight rather than fight, betting that the better horses of the express would allow the riders to outdistance attackers.

The first electric words flashed across the country via the telegraph on October 24, 1861, and the pony express quickly went out of existence. The country marveled at how a message that had of late taken ten days to cross the West now took only the time necessary to tap it out in Morse code.

During the summer of 1866, the Platte valley echoed with the sound of the Pacific railroad pushing westward—the ring of spike mauls on iron, the rough talk of Irish trackbuilders, the hiss of steam. Mile by mile the Union Pacific pushed west, following the easy grade of the valley toward the summit in Wyoming and beyond to the mountains of Utah. Meanwhile, the Central Pacific struggled with the rock of the Sierras as it built eastward to the eventual meeting and the golden spike.

Just as it had been for those who came before, the Platte River valley was the perfect route. The steel rails curved gently to follow the broad S curve of the valley as they made their way across Nebraska. This was the bold expression of the white man's conquest of North America. Though the frontier would not be considered closed until 1890, the land of the red man was divided and set for conquest in 1869 with the golden spike, where East met West on May 10 at Promontory, Utah.

The broad Platte valley had become a pathway for a striking number and variety of travelers, machines, and communication devices. Everything but steamboats adopted it because of its ideal location, its clear superiority to any other nearby routes, and the fact that the river was only a few inches deep in midsummer. The towns in the valley all owe their beginnings to transportation; most were stations and division points platted and named by the Union Pacific as it built west. They grew into farm and ranch centers, but their polarized heritage along the line of the river and the line of travel is still very apparent to anyone who looks at a map.

Lincoln Highway travelers entered the West through a corridor that blunted the transition, a protected river valley that made things seem a bit like Iowa when in fact they had become more like Wyoming. Although the country grew dryer and the land began to stretch apart, travelers were unaware of just how great the change actually was. Wetland farming continued along the Platte far west of where the uplands had given over to rangeland and wheat. Few travelers took the time to turn out and drive up onto the plains to observe how the country had changed. Instead they pressed on and looked westward for their first view of a butte or snowcapped mountain. Between Big Springs and Chappell the highway turned out of the South Platte valley and climbed to the plains for a distance until it dropped to follow Lodgepole Creek. Here on the high plains Bellamy Partridge discovered just how the land had changed as he had made his way west in the valley:

We had just splashed through the mudholes at a place called Big Springs when we ran up a longish grade and came out on the top of a tableland where for the first time we found ourselves on the plains—the western prairies of song and story.

As I looked all around and saw nothing but space—and so much of it— a feeling of loneliness came over me. Without realizing what I was doing I stopped the car and shut off the engine.

I listened intently for the music of the spheres, the whispering of the stars passing overhead in the daytime, the swishing of the planets through the immeasurable nothingness—but I couldn't hear a thing.

Only five miles from where Partridge stopped the car and looked about in awed silence, the Oregon- and California-bound pioneers made a similar discovery when at last they climbed out of the valley and traveled across the Great Plains for the first time. They, too, had traveled for endless days in a seemingly unchanging valley, wondering if they ever would get to the West. Then, 490 miles out from Independence, at a point just west of present-day Brule, the main trail crossed the South Platte River and turned out of the valley northwestward and climbed the hills to the high plains. Though no buttes or mountains were in sight, the overlanders knew that they were at last in the West; they knew that very little of what they would see from then on would remind them much of Missouri, Iowa, or home. No doubt, like Partridge, many wagons pulled to a halt as pioneers scanned the horizon around them. No longer hemmed in by hills, they must have felt they could see clear to California. Even though there remained some 1,500 miles yet to go, in a burst of optimism they named the place California Hill.

After the completion of the transcontinental railroad in 1869, the trail fell to disuse. The railroad had taken away the stagecoach and the overland wagon, and much of the historic path returned to grassland or was plowed under for crops. In 1906 Vera Marie Teape came west by automobile, making the best trail she could, using parts of the old wagon trail and parts of what would later become the Lincoln Highway. On the rutted old trail where fifty years earlier some 300,000 people had passed on foot, wagon, and horseback, she found sunflowers thick and high enough to break the drive chain on her car.

The Lincoln Highway never followed the exact ruts of the Oregon and California Trail. The Lincoln stayed on the north side of the Platte, staying close to the Union Pacific and passing through the towns that the railroad brought. The main route of travel for the wagon trains had generally been along the south bank.

But at California Hill the Lincoln crossed a main route of the old trail, the Lincoln Highway pressing straight west, the old trail turning out of the valley to the northwest. The faint ruts made by the wagons climbing out of the valley were visible to Lincoln Highway travelers as they drove to the exposition in 1915, as they still are today. But little other evidence of this great migration, the pony express, the overland stages, or the trappers survived to remind them of the great role this path had once played in settling the West. To be sure, the bright rails of the Union Pacific were always nearby, but of the earlier and more colorful days little remained.

Soon after entering Wyoming, the Lincoln Highway left the valley of Lodgepole Creek—hardly a valley at all this far upstream—as the stream disappeared off to the north and the Lincoln Highway rose over the high plains and dropped toward Cheyenne. The mountains were first seen west of Burns, Wyoming, only twenty-six miles from Cheyenne. Unlike a traveler approaching Denver, who could see the long-awaited Rockies for a hundred miles or more, the Lincoln Highway traveler was almost upon them before they appeared.

Cheyenne to Salt Lake City—476 Miles

Via Laramie, Rawlins, Rock Springs, and Evanston. The Lincoln Highway takes the auto tourist over Sherman Summit, across the Laramie Plains, the rough country at the north end of the Medicine Bow Range, and the Great Divide Basin. The road passes buttes and rimrock as it approaches Green River, dropping into Echo Canyon just beyond Evanston. Echo, Weber, and Parleys canyons then see the traveler to Salt Lake City.

Cheyenne, like Chicago, was a reprovision point along the early highway. Travelers replaced tires and tubes and bought equipment intended for emergency use in this dry country—canvas water bags, extra gas cans, and a supply of food. For the first time the early Lincoln Highway travelers would be away from civilization; repairs, hotels, and hot meals would be scarce from here to California.

West of Cheyenne, the nature of the terrain and the road changed abruptly. Travelers left the gentle rolling plains and wheatfields of eastern Wyoming and began a long ascent up the face of the Laramie Mountains. They had entered the western mountains at last as they climbed from Cheyenne's 6,062-foot elevation to Sherman Summit, 8,835 feet above sea level. This was the highest point Lincoln Highway travelers reached, though it was not the continental divide. Sherman Summit was 1,700 feet higher than the next highest mountain crossing on the route: Donner Pass in the Sierras of California. Rather than the sylvan glens and cold mountain streams that tourists expected and that Donner would later provide, Sherman was really more of an uplifted arid plain, broken by rock outcrops and valleys. Trees were few, and high winds pulled at the thin grass. The alpine forests and rushing streams would have to wait until California.

Although many were puzzled at the dry nature of the place, travelers were awed by the view. The Lincoln Highway topped the mountains not far from where the Union Pacific tracks crossed the summit, and like thousands of train travelers before them, the auto tourists were taken in by the wide vistas of mountains in the distance. To the south were the high peaks of Colorado, to the west the Medicine Bow Range and Elk Mountain. Here the distant mountains weren't seen from below, as from a valley; on Sherman Summit the auto tourist stopped and viewed the peaks at shoulder height through the thin, clear air.

With the change in landform came a great change in the road. From here all the way to Evanston at the western border of the state the men and women who drove the early highway followed a comparatively new route; it was no longer a well-trampled valley where paths and conveyances of all sorts had led the way. The highway did stay close to its Nebraska traveling companion, the Union Pacific Railroad, but prior to its construction across the southern part of Wyoming, most of this route had had little transportation significance. While it was considered to be more or less part of the great South Pass route, that trail actually looped a hundred miles to the north, and the emigrants, the pony express, and the stage route with it. As a result, there had been no real road of any importance here along the railroad. Like the roads of the Midwest, this diffuse path was used by local travelers who knew the way from ranch or mine to town, but before the auto, it had never been traveled by strangers.

So little was known about the route west of Laramie that when Bellamy Partridge arrived there in 1913, the only guide to the road on to Utah was a hand-drawn map made on an earlier trip by a Mr. Lovejoy, a Laramie resident. He loaned it to Partridge with the agreement that it was to be updated and mailed back from the western border of the state.

The early road crossed Wyoming with a series of meanders. Though the route generally struck fairly close to the tracks, depending on season, local advice, and pure whim, the traveler might find himself or herself on one of several paths. In many places auto travelers simply went overland, following the terrain through arroyos and irrigation ditches, opening and closing gates and following a trail as best they could. The road wound across the high plains, diving into ravines and veering around rock outcrops, wandering through the gray sagebrush. Year by year this route between Cheyenne and Evanston would evolve, travelers would hunt and search for a better way across the state, and like an electrical current, they would seek the path of least resistance.

Alice Ramsey, who traveled to San Francisco in 1909, found circumstances that were little different from what motorists would find four years later when the Lincoln Highway was established, or even in 1915:

Roads in Wyoming were scarcely what we would designate as such; they were wagon trails, pure and simple; at times, mere horse trails. Where the conveyances had usually been drawn by a team, there would be just the two definite tracks—or maybe ruts—often grass-grown in between. On the other hand, where many one-horse rigs had passed, a third track would be visible in the middle through the grass or weeds. With no signboards and not too many telegraph poles, it was an easy matter to pick up a side trail and find oneself arrived at the wrong destination.

Bridges were another problem in the early days. No matter how steep the sides, no matter how rough the streambed, a dry arroyo or ravine was seldom honored with a bridge. Hours could be spent cutting down side banks, moving rocks, and wrestling heavy cars with primitive brakes and transmissions through ravines with nearly vertical sides. If a traveler was unlucky enough to make a trip across shortly after a Wyoming cloudburst, he or she would often find the work of previous travelers washed away and would have to start anew. Most of the more major stream crossings had bridges, but even they would be washed out periodically in the locally heavy rains. When the water was high or the bridge washed out, the traveler had few choices. One could wait for the water to go down, drive up- or downstream in hopes of finding another bridge, or, as most travelers did, use the railroad bridge.

Those who chose this option coordinated their crossing with the busy Union Pacific and drove over during times when no trains were due. Small telegraph stations were common along the line in those days, and the railroad was cooperative. No doubt there were a few who didn't bother to alert the railroad, a few who took their chances, looking nervously in both directions, probably putting an ear to the rail to listen, and hurrying across on the narrow spans. As they slowly bumped over the ties, they couldn't know whether they'd be killed by tipping off into the stream or by colliding with an oncoming train.

Beyond Sherman Summit and Laramie, the 1915 road turned northwest and crossed the Laramie Plains, a wide valley crisscrossed by irrigation ditches and scattered with cattle ranches. During times of wet weather or heavy irrigation, travelers became mired in the soft soil or sometimes wandered about in a maze of irrigation ditches seeking an outlet. To the west were the peaks of the Medicine Bow Range, to the east the undulating ridge of the Laramie Mountains.

Beyond Rock River, the country became quite broken as the road angled back west and crossed the tag end of the Medicine Bow Range. Leading across plateau and ridge, and through stream valleys, the highway picked its way through Medicine Bow and Hanna to Rawlins.

Wyoming was where travelers crossed the backbone of the continent. It was not a spine—a sharp, abrupt range of mountains—but a high and broad plateau that projected the auto travelers against the sky. Here the land rose up close to the sun and clouds on the arch of the continent. The sky pressed low on days of overcast; on sunny days, the traveler in an open car felt like a part of the ocean of air above. In Wyoming the true West proved itself in the limitless vistas of sky, sage, buttes, and distant mountains. The clear, thin air enhanced the blue of the sky and the sharpness of the clouds, and made all the colors in the landscape vivid. The horizon was clear and distinct at great distance, with snow-clad peaks far to the north and south, blue mountains that could be seen for hours as travelers passed slowly across the broad land. Closer by were buttes and rims, glowing brown, red, and yellow in the bright light. All around was endless grey-green sage, dusty and fragrant on the rocky soil.

Twenty-four miles west of Rawlins, the Lincoln Highway crossed the great continental divide, marked by a small sign. At last the autoists had attained the lofty pinnacle, the ridgepole of the continent— but it was nothing more than a nondescript ridge no more impressive, and in fact quite a bit lower, than some they had crossed earlier. This was just another sage- and rock-covered hill like so many others. At

7,178 feet it was 1,657 feet lower than Sherman Summit, which certainly looked a lot more like a continental divide than this place. Most travelers were surprised, and many were confused. Even the Lincoln Highway Association had it wrong. The 1916 guide told the traveler that Sherman Summit was the line between Atlantic- and Pacific-bound waters. But here was the true place every traveler anticipated, marked only by a small sign that mirrored the anticlimax of the terrain and the event.

This was actually the beginning of the Great Divide Basin, a vast undrained area measuring some sixty by one hundred miles. The Lincoln Highway edged along the south ridge of the basin; occasional views of endless sand and scant vegetation to the north no doubt made travelers thirsty even if the strong sun and dry air did not. The area was also known as Red Desert and was received with scant enthusiasm by some overland travelers. One auto nomad, listing his high points for the day, noted the crossing of "Red Desert, and what we called Yellow Desert and White Desert, and many other deserts, which all looked much alike to me—being simply deserts—where even the prairie dogs, rattlesnakes and coyotes had given up the country in disgust."

West of Rawlins, the earliest Lincoln Highway followed an old grade of the Union Pacific most of the way to Wamsutter. The railroad embankment was narrow, which made meeting another car an interesting experience, but in general it made a good road because the grades and curves were gentle and the jumbled landscape of ravines and rock could be crossed on smooth fill. The frequent trains of the Union Pacific passed nearby on the newer, straighter alignment.

The Lincoln Highway moved close to the railroad near Point of Rocks, where the road entered the valley of Bitter Creek. Here again was the trail of the overland stages, last seen near Chappell, Nebraska. It had been moved south in 1862 to a route somewhat along the Colorado-Wyoming border and away from the South Pass route be-cause of Indian raids. The Lincoln descended in the dust of the old stage route to Green River—both the town and the stream—as the high rim or scarp rose on either side. Beyond the river, the highway climbed once more to high, open plain for the last few miles of plateau, for a last look from the roofline of the continent.

Early Lincoln Highway travelers reacted to Wyoming with more emotion than to most other places. Most were fascinated by this landscape that changed so greatly from hour to hour across the state—especially westbound travelers who had spent days following the Platte valley across Nebraska. Wyoming impressions were as vivid and distinctive as her land and sky. As a rule, motorists loved "the tawny and dusty olive of houseless land running away to the horizon," but a few found the landscape dull. One autoist, a malcontent who complained about almost everything, found Wyoming "about as scenically stupid as Kansas," which says little for either state. For another, all it took was the scent of sage, or the similar smell of crushed daisy stems, to bring to mind "the sharp, untempered angles of the buttes, clear in that dustless, diamond-bright air and to feel the cool vigor of the west wind, coming down from snowclad peaks."

Nightfall, however, sometimes brought a change of heart to the traveler from the urban East. Thornton Round and his family, returning to Cleveland in 1914, collapsed the front wheels of their Model T in the sagebrush east of Rock Springs. They summoned help but it was a long time arriving. As evening turned to night, their city hearts beat faster:

Dad and I settled down in the front seat of the car and tried to relax a bit. This was hard to do though. We were surrounded by the blackest of nights and some of the queerest noises I have ever heard. A blood curdling scream came from somewhere out on the prairie. This was followed by a loud cry that might have been made by a loon or some other large bird. A coyote joined the eerie chorus. Several road runners ran around the car, and as they moved they gave funny little squeals. Whippoorwills called, and we fancied that we could hear a wolf baying. Nature can sometimes be terrifying as well as awe-inspiring.

Later, help arrived; the car was repaired and the Rounds spent the rest of the night safe in a hotel in Rock Springs. "I can tell you that any bed would have satisfied us that night, just as long as it wasn't out in the middle of the desert." The next morning, they passed the scene of their nocturnal fright: "Everything looked serene and peaceful. There was not a sign of a wild animal or a noisy bird."

At Granger, thirty-four miles west of Green River, modern U.S. 30 splits from the Lincoln Highway and heads northwest to Idaho and Oregon, ending at the mouth of the Columbia River at Astoria. The Lincoln Highway and U.S. 30 had shared a common route for 2,191 miles, beginning in the heart of Philadelphia and ending at a tiny railroad town on the high plateau of Wyoming.

As U.S. 30 splits off at Granger, the main routes of the Oregon and California trails rejoin the Lincoln Highway for the thirty-six miles to old Fort Bridger, an early trading post on the trail. The major routes of these trails then turn north to Fort Hall in Idaho, while the Mormon Trail to Salt Lake runs for many miles with the Lincoln Highway. The pony express route also rejoins the Lincoln at Granger, but unlike the covered wagon trails, its connection with the Lincoln Highway will continue most of the way to California.

A few short miles west of Evanston, the Lincoln Highway leaves Wyoming and at the same time leaves the high land of open vistas for a narrow path in a deep canyon. At a point six miles into Utah, the highway tips downward at the head of Echo Canyon, and a long descent begins. From Evanston to Echo, Utah, the road drops 1,300 feet in thirty-two miles with a long, steady downgrade along Echo Canyon Creek. The rails of the Union Pacific twist back and forth alongside,

trying to increase the distance in order to lessen the grade per mile. Tourists in trains and autos alike craned their necks to see Steamboat Rock, Giant Teapot, Sphinx, Gibraltar, Sentinel, the Cathedral, and Pulpit Rock.

The Lincoln Highway leaves the main line of the railroad and turns south and upstream along the Weber River near the town of Echo. Turning west again, the road crosses Parleys Summit and enters Salt Lake City through Parleys Canyon. At Salt Lake City, the downward slant of the road ends; the terrain will rise again to the west. Here at the edge of the Great Basin, the Lincoln Highway is a half mile lower than it was back on the windswept landscape near Evanston.

As westbound Lincoln Highway travelers emerged from Parleys Canyon, they could see the Great Salt Lake and a bit of the Great Salt Lake Desert beyond the city. These features dominated the horizon as a vast, table-flat expanse of white sand, salt, and water. Even as travelers entered the capital city of Utah, their eyes must have lifted to the distant line of white that shimmered in the summer heat, broken now and then by dry mountains. They would have thought of pictures they'd seen of abandoned wagons and the bones of oxen, all bleached white in the relentless sun, left far out in the desert where the oxen had died of thirst. They could just as easily imagine broken automobiles scattered across that disorienting waste.

As they sought comfortable hotels along the wide, tree-lined streets of Salt Lake City and reveled in city amenities once more, they must have grown anxious about what was to come, about the challenge they would soon face to the west. They knew they had accomplished much, but the final and greatest test was to come. Wyoming had been dry, and many described it as a desert, but the Great Salt Lake Desert caused any to pale in comparison, and must have caused many a tourist to pale in contemplation of crossing it.

6

Salt Lake City to San Francisco: Desert, Mountain, and Sea

Imagine a vast, waveless ocean stricken dead and turned to ashes; imagine this solemn waste tufted with ash-dusted sagebushes; imagine the lifeless silence and solitude that belong to such a place; imagine a coach, creeping like a bug through the midst of this shoreless level, and sending up tumbled volumes of dust as if it were a bug that went by steam; imagine this aching monotony of toiling and plowing kept up hour after hour, and the shore still as far away as ever, apparently; imagine team, driver, coach and passengers so deeply coated with ashes that they are all one colorless color; imagine ash-drifts roosting above moustaches and eyebrows like snow accumulations on boughs and bushes. This is the reality of it.
  —Mark Twain,
      Roughing It

Salt Lake City to Ely—287 Miles

Via the Great Salt Lake Desert through Fish Springs, Ibapah, Tippett Ranch. Only scattered ranches and tiny settlements are passed crossing this expanse of white desert, dry mountains, and uninhabited valleys. The traveler is far from civilization for nearly all of the distance and should be prepared with extra water, foodstuffs, and repair parts. The Lincoln Highway turns from the shore of the Great Salt Lake and follows the edge of the salt desert, then climbs and descends through mountainous terrain to emerge in Steptoe Valley near Ely.

The 1915 Lincoln Highway struck due west leaving Salt Lake City, toward the south shore of the lake. The city ended rather abruptly and desert landscape, a sort of scenery that would soon become very familiar, took its place. Salt Lake City lay safely tucked against the Wasatch Range behind, as if afraid to approach or enter the white desert. The highway passed the famous Saltaire resort built on stilts over the lake, skirted the edge of the Salt Lake for a time, and then turned south into Skull Valley.

In that baked valley travelers finally faced the enormity of the task ahead. Cold dread must have arrived to become another passenger, to sit silently on the water cans, spare parts, and extra tires. This valley was stark, dry and uninhabited, and introduced what was ahead. Beyond the south end of Skull Valley lay the Great Salt Lake Desert, the very pit of the great interior basin of the western United States, an alien landscape of burning salt and sand, devoid of potable water. Travelers thought of the treacherous mud flats waiting to trap automobiles, dry washes where springs could snap with ease, rough mountain-pass roads with sharp stones galore to slash tires, thick alkali dust to choke carburetors and throats alike, and most of all, heat; heat enough to boil the most efficient radiator, dehydrate the stranded traveler, and deplete the fullest water bags.

This was difficult country for automobile tourists. Joy and his associates had worked hard to select a route that provided the best access to water and habitation and that avoided the worst terrain. The choices hadn't been easy.

When Joy first came looking for the Lincoln Highway route, he had found three possible auto routes between Salt Lake City and Reno, some 530 miles. The first went north from Salt Lake City and became a crude trail beyond Ogden, where it looped around the north end of the Great Salt Lake and the salt desert. It roughly followed the old line of the Central Pacific Railroad—later Southern Pacific, the western half of the transcontinental railway. Few people lived north of the lake, leaving the traveler in dire straits in case of trouble. This was also the least direct route around the lake and desert barrier. It entered Nevada near Montello and came to follow the Humboldt River— the path of the old California Trail and the Southern Pacific and Western Pacific railroads. The Humboldt River valley was a corridor of travel much like the Platte in Nebraska, and auto travelers on this route passed through many railroad towns, including Elko and Winnemucca, before reaching Reno.

The second route between Salt Lake City and Reno, a central route, was the most direct. It went straight west from Salt Lake City, along the south edge of the lake and directly across the middle of the salt desert. It entered Nevada just beyond Wendover, and like the north trail it followed the Humboldt valley route to Reno. Interstate 80 closely follows this route today. This was by far the most direct path, but was a perilous choice at best. It challenged the desert head-on and required driving some forty miles of stark bone-white salt flats, a true speedway in places even when wet, but in others it turned to the most tenacious mud with a little rainfall. Only the old desert men seemed to be able to tell which was which.

In 1848, a pack train of mules led by an experienced western traveler nearly came to grief making a direct crossing of the salt desert.

The packers were following the trail of an earlier group which had crossed quickly and uneventfully a few years before. When well out onto the salt flats, the mule train was overtaken by a cloudburst and the animals immediately sank into the salt mud. The packers abandoned all provisions, equipment, and saddles in an attempt to save themselves and the mules. They turned back, and after forty-eight hours of exhausting struggle without food or fresh water, they escaped with their lives and little else.

Last was a southern route, the route that Henry Joy selected for the Lincoln Highway. It went more or less southwest from Salt Lake City and skirted the south edge of the worst salt desert. The first place of import it touched in Nevada was Ely, and from there it went west through Eureka and Austin to Reno. At Ely, good connections could be made for Los Angeles via the Midland Trail. Though it was more direct than the north route, and much safer than attempting to cross the salt desert directly on the middle route, this path between Salt Lake City and Reno was by no means a straight line or any sort of speedway. It was a path of broad desert and numerous obstructive mountain ranges.

This south route was an established path and had been the route of the pony express, the overland stage, and the transcontinental telegraph. Though the Lincoln Highway didn't follow those routes in their entirety, it ran on the trail of the stages and express riders for most of the distance through the hardest desert. Joy selected this route to California around the south perimeter of the salt desert for the same reasons that the scouts for those operations had—water could be found at intervals, the route was passable and direct as any route could be, and it avoided the worst of the salt flats.

The most crooked and the roughest part was between Salt Lake City and Ely. As the crow flies, it is 185 miles between those points; via the 1915 Lincoln Highway, it was 287 miles. For some distance the route followed the scalloped edge of the salt flats along the shoreline of the ancient and once larger lake. It turned this way and that; it braved a few parts of the open desert, crossed or skirted seven ridges or small mountain ranges, and inched through an equal number of rocky, dry valleys between. There were some long stretches of sand and a section or two of the much-dreaded salt flats. The road was rough, necessitating slow travel, making the travel time between water stops agonizingly long.

Between Salt Lake City and Ely this route passed through nearly uninhabited desert. There would be no railroad to follow as across Wyoming, no more ranches and towns scattered across the plain; here a ranch or tiny settlement was a major landmark. They would be seen for many miles and anticipated for many before. The only place that could be considered a town was Grantsville, forty miles west of Salt Lake, where there were three hotels but no garage. The guidebooks and maps listed the ranches as they listed towns elsewhere: Brown's Ranch, Orr's Ranch, and the landmark J. J. Thomas Ranch at Fish Springs. Under the listing for Fish Springs, the Lincoln Highway guide for 1916 said: "Ranch meals and lodging. Hot sulphur springs close to ranch. If trouble is experienced, build a sage brush fire. Mr. Thomas will come with a team. He can see you 20 miles off."

Most early travelers boarded and reprovisioned at the ranches along the way. Ranch men and women—even the hermit Thomas—were generally hospitable to travelers who might drop in at any time of day or night. Autoists seldom intended to drive after dark in this country, but delays were common, and the arrival of a car full of hungry tourists in the wee hours was not uncommon, nor were ranchers likely to be upset by the event, especially when it meant a good bit of pocket change. A ranch family could keep extra foodstuffs and supplies on hand and expect to get good return on their investment. The tourist business often supplemented a meager ranch income and provided a pleasant break from an isolated existence.

Ironically, the greatest hazard to travel in this arid country was water, just as the well-known Thomas mudhole suggests. There were many constantly wet places in the desert, and though rain was infrequent, when it came it was usually in cloudburst form. It could be dry for months at a time, then rain a couple of inches in an hour. Since the streambeds and ravines were nearly always dry, there were few bridges across them. The traveler would simply scramble down the bank, across the bed, and up the other side. But a quick storm across the desert instantly turned these arroyos into torrents of muddy water. The flood usually subsided quickly, but the erosion of the bed sometimes left the road impassable.

The road crossed a few sections of salt flat and salt marsh, and the condition of these places was of great concern to long-distance travelers from the green East, where the habits of water were predictable and its presence apparent. It took only a few hundred feet of sticky salt mud to make the entire trip impossible. This south route avoided crossing any broad expanses of the flats, but what little there was could trap a car just as effectively. Hard as pavement when dry, an inch of rain would turn them to a mud so soft it could absorb even a stationary auto.

From the southern end of Skull Valley, near the present site of Dugway, the 1915 Lincoln Highway struck south-southwest toward Granite Peak across rocky desert country. Not far from County Well ("Just a well. Radiator water only," warned the guide), the road passed between Granite Peak and the north end of the Dugway Range, and the salt desert loomed ahead. To the north, it stretched endlessly toward faint mountains shimmering in the distance.

It was somewhere to the northwest that the Donner-Reed party of California-bound emigrants in 1846 crossed the horizontal whiteness behind plodding oxen. Like the pack train that crossed two years later, they became deeply mired in the muck. Oxen struggled to free the heavy wagons but soon fell with thirst and exhaustion. The emigrants became desperate and abandoned possessions, wagons, and finally, all hope. They avoided disaster only because some were able to push ahead and return with water to revive many who had resigned themselves to death on the desert. They survived only to meet true disaster later in the mountains of California.

The Lincoln Highway angled south, away from the white sea of salt flat and stayed close to the Dugway Range to the east. At the north tip of the Black Rock Hills, the traveler passed a dark volcanic outcrop of rock and turned west. Here the Lincoln Highway joined the routes of the pony express and the overland stage. The Blackrock pony express station once stood at this point.

Here was a road infused with history. The narrow tires of Maxwells and Fords ran in the very ruts made by the overland stages and in the hoof tracks of the pony express. In fact, the stage still ran along this line as late as 1909. It was an unimproved trail, following the terrain, connecting one waterhole with another. The path had changed little if at all since the time of the first regular travel over this route about 1860. The ruins of several stage and pony express stations were found at ten-mile intervals. Given a mochila and a fast horse, an old pony express rider could have easily ridden the route and would have noticed little change with the passage of fifty years.

A few miles short of Fish Springs, the road crossed two miles of mud flats and salt marsh which gave early travelers a small sample of what crossing the salt desert would have been like. One group of auto travelers stopped here for a picture on a hot and fortunately dry day. Their photo was captioned "Death desert—100 miles to water, 3 feet to hades." Less fortunate travelers on wetter days built sagebrush fires summoning Mr. Thomas and his team to pull out their mired autos. Once past the mud, the road soon passed the old pony express station at Fish Springs where Thomas lived the hermit's life.

The road angled around the north end of the Fish Springs Range, staying close to the rocks and overlooking the flat and endless desert to the north, then struck west across open desert toward Callao, fifteen

miles away and clearly visible as an oasis against the high, snowcapped Deep Creek Mountains beyond. The awful desert would soon be behind. Northbound along the west edge of the salt, the road climbed quickly along the face of the Deep Creek Range, then turned and entered the mountains through a narrow canyon. Looking back, one could glimpse the line of the white desert for the last time as the road twisted along the rising streambed.

Once beyond the salt desert, travelers began to relax a bit. There was still plenty of hard going ahead, but the specter of dying of thirst on that nightmarish expanse no longer haunted them. Somehow California seemed much closer just for having the worst of the desert behind.

Most overland auto travelers had surprisingly little trouble on the hard desert. They broke a few springs, got stuck in the mud flats from time to time, but no one seems to have suffered any serious deprivations or even much delay in crossing. Most had been sufficiently warned of the dangers and so were well prepared with extra water and supplies, and their cars were put in tip-top shape before the crossing. Their fear caused them to travel so cautiously that mishaps were few.

Though the salt desert caused the greatest worry to travelers, back in Iowa was where they had the most trouble. More travelers were probably delayed by Iowa mud and high water in the wet year of 1915 than by all the cloudbursts, washouts, and deep salt mire during the twelve years that the Lincoln Highway used this route across the desert southwest of Salt Lake City. The danger of the desert was greater, but the real hazard to travel lay in the soft hills of corn and hay in Iowa.

The road ascended the Deep Creek Range, dropped to Deep Creek Valley, and passed through Ibapah; it crossed sage desert and entered Nevada without fanfare. It angled across a broad valley, where Tippett Ranch could be seen for miles in either direction, passed by the south end of the Antelope Range, crossed Spring Valley, ascended the Shell

Creek Range, and at last dropped sharply to Steptoe Valley. Near the faint markings of a pony express station there, the Lincoln Highway turned south toward Ely, the first real town since Salt Lake City.

Ely to Reno—336 Miles

Via Eureka, Austin, and Fallon. The traveler crosses the heart of Nevada by ascending and descending several abrupt mountain ranges alternating with wide, unobstructed valleys. The landscape is primarily desertlike, except for near the tops of the mountain passes, where thin pine forest is found.

The climb over the Deep Creek Range into Nevada had not only signaled the last of the hard desert, but it also announced a new topography that would dominate the landscape until the Sierra wall was attained in California. The beginnings of it had been observable even in the Utah desert—most mountain ranges stood like compass needles, pointing directly north and south, separated by wide valleys. The Stansbury and Cedar mountains had formed Skull Valley, a sort of proscenium arch leading the traveler to a new topography. This is the geography and geology of basin and range.

From the Deep Creek Range clear to California, the mountain ranges of Utah and Nevada line up north and south like great swells in an ocean of stone. Each range is a sharp ridge or a single line of peaks, separated from the next by a broad valley. Though narrow, these ranges are dramatic when set behind the wide, unobstructed valleys. The Lincoln Highway traveler approached each wave at right angles, crossed valley floors at about six thousand feet above sea level, climbed to passes at seven thousand feet, and descended to the next valley. Many peaks in these ranges overlook the valleys from ten thousand feet.

Even Effie Gladding, who was hardly the most astute observer of the landscape, could not help but notice the topography:

We were passing from one great valley into another, hour after hour. When I looked on the map of Nevada, I found a series of short mountain ranges. I could see what we were doing in our travel. We were descending into a valley, crossing its immense width, coming up on to a more or less lofty pass, usually bare, and descending into another valley. It was very fascinating, this rising and falling with always the new vista of a new valley just opening before us.

Striking west from Ely headed for Fallon, the traveler knew that only six hundred miles remained to San Francisco; better than 2,700 lay in the cloud of dust behind. This part of the Lincoln Highway became U.S. 50 in 1925.

In the early days of the Lincoln Highway, this route, like many through the West, was difficult to follow. It had been assembled from a network of informal wagon roads, and its identity as the Lincoln Highway depended on signs and banded telephone poles which pointed the way when a fork was encountered. Should there be no pole line, or should the sign have blown down or been removed, the traveler was often hard pressed to select the correct fork. Because of heavy wagon freighting to and from the many mines scattered across the state, the through road may well not be the more traveled one. Heavy mine traffic was common, but transcontinental motorists were few. Occasionally travelers would make an unintended call at a ranch or mine when what they thought was the right road suddenly ended in a dooryard or at a mine headshaft.

The Lincoln Highway route across the central part of Nevada was much newer and far less known than a parallel route through Elko and Winnemucca along the Humboldt River, ninety miles to the north.

Because of available grass and water, the north route had been the path of the California Trail, and because of the flatter terrain, it became the route of the Southern Pacific and Western Pacific railroads. It was a corridor, not unlike the Platte River valley in Nebraska, where towns had sprung up along the course of travel, and although the road itself was very poor, it was strong competition for the Lincoln in the matter of auto traffic. Westbound travelers crossing Wyoming often debated the relative merits of taking the Lincoln around the south side of the salt lake and desert, then crossing Nevada over basin and range, or taking the north route around the lake and following the Humboldt River route to Reno.

Henry Joy selected the south route across central Nevada because it lined up with the route around the Great Salt Lake Desert. After swinging south around the salt desert of Utah, his direct route could not turn back north to Wells on the Humboldt River, adding better than a hundred miles to the distance from Salt Lake City to Reno. Instead, he pointed it straight toward Reno, across the ranges of Nevada. This south route also made it easy to make a connection for Los Angeles at Ely. Besides, Joy thought this central route was a good one. A Packard could climb grades that a locomotive could not, and grass meant nothing to a tireless internal-combustion engine.

First out of Ely was Robinson Summit, then Little Antelope Summit, Pancake Summit, Pinto Summit, and the town of Eureka. Between there and Austin was Hickison Summit, which for many years was also known as "Ford's Defeat," owing to the numbers of Model Ts which burned up transmission bands ascending the steep grade and had to be pulled to the top. A long downhill coast brought them to the foot of Scott and Austin summits. Although a defeated Ford was only fifteen miles from Austin, these two high passes stood between the motorist and repair.

Between Ely and Fallon, 270 miles, the only towns were Eureka and Austin. Although they were the most remote towns that Lincoln Highway travelers visited on the cross-continent journey, both boasted hotels and garages and gave the traveler a taste of civilization in this wide land of sage, rock, and sky. Utah settlements like Callao and Ibapah certainly were more isolated, but they were hardly towns. Eureka was a county seat, and both Eureka and Austin were important supply centers for ranches and mines for a large part of central Nevada. Except for the new Lincoln Highway, they stood many mountain ranges and valleys away from the pulse of modern life. Weedy railroad branch lines reached south through valleys to connect the towns with the high iron along the Humboldt, and wagon roads came in from everywhere; but still these towns stood remote, across nearly one hundred miles of sagebrush, far from the outside world.

Eureka and Austin had both begun life as silver boom towns, but by the time the Lincoln Highway came through, the ore had played out and only a few mines and smelters struggled along extracting silver and lead from the hard ore. The businesses in both towns catered to a mix of miners, ranchers, cowboys, and now, with the highway, a flurry of automobile tourists. The interactions between "outlanders" and locals in bars, hotels, and stores were seldom unpleasant, and travelers often remembered them. In 1913, the year the Lincoln Highway was announced, Bellamy Partridge and his party of travelers approached Eureka and passed many men on horseback, all dressed up and "wearing gay kerchiefs and fancy shirts." Upon reaching town, they discovered that there was to be a dance that night:

Music was furnished by a cowboy band, the leader of which explained to me that since they never had any time to practice together they had to learn the pieces separately and were just a group of soloists. The band had plenty of volume and kept very good time though I suspected once or twice that most of them were playing different tunes.

I tried to get the girls to go to the ball, but after listening to the music, they went to bed instead. Personally I think they missed the chance of a lifetime. They would have had at least ten partners apiece for every dance.

A day later they were in Austin, where they sat in on a political rally for Teddy Roosevelt and the Roughriders.

Beyond Austin, westbound travelers crossed Railroad Pass, drove around aptly named Dry Lake, ascended Carroll Summit, passed— and in the early days, opened and closed—Eastgate, Middle Gate, and Westgate, and just beyond Frenchman crossed a low pass, where they left the desert and dry mountains behind. Ahead lay a bad stretch of sand and alkali flat, but beyond was the wide valley around Fallon where irrigation brought crops, grass, and trees to the land.

In true comic fashion Bellamy Partridge got lost among the irrigation ditches near Fallon. After great care had saved him and his wife from disorientation crossing the vast ridges and valleys to the east, they lost the trail where civilization had altered the landscape:

Beyond the Sink the roads spread out like a fan in the soft sands of a huge irrigation project. I stopped at the cabin of a forest ranger to ask directions. He gave them with numerous turnings right and left to take us over the ditches. We went on for a while but did not seem to be getting anywhere. Then I happened to see another ranger's cabin. I drew up in front—and out came the same man. There were no signs of recognition between us. I asked the same question. He gave the same answers, being too well mannered to humiliate me.

I would probably have driven away and, likely as not, would have come back again. But my wife leaned out and handed him an envelope. "Would you please write it down?" she asked pleasantly.

He did, and by following his notations closely, we were extricated from the maze and reached Fallon for our overnight stop.

Nine miles west of Fallon, the route of the Lincoln Highway split into two sections. The main route continued west through Reno, over Donner Pass, through Auburn to Sacramento. The longer Pioneer Branch turned somewhat more southwest through Carson City and around the south end of Lake Tahoe, crossed the Sierras over Echo Summit, went through Placerville, and rejoined the main route at Sacramento. In later years the Pioneer Branch became U.S. 50 and the Donner Pass route became U.S. 40. Interstate 80 follows the northern route today.

Beyond Fallon, the main route of the Lincoln Highway snaked for some twenty miles through Truckee Canyon, a narrow gash in the Virginia Mountains, then suddenly emerged into Truckee Meadows, the site of Sparks and Reno.

Reno to Sacramento—143 Miles

Via Truckee, Donner Pass, Emigrant Gap, Auburn. West of Reno, the Lincoln makes a sharp climb to the crest of the Sierras at Donner Pass, then a long descent to the great valley of California at Sacramento. The geography changes from high desert in Nevada to alpine crags in the Sierras, to dense forest on the western slopes, to palm trees and agriculture in the flat valley.

In the fall of 1846, the men, women, and children of the Donner-Reed company struggled through Truckee Canyon, crossing and recrossing the swift, cold Truckee River at least twenty times with their oxen and heavy wagons. The ordeal in the desert and the circuitous route they followed to get to this point had made them late in approaching the last and greatest barrier to their safe California arrival— the high Sierra Nevada. Many other emigrants had crossed earlier that year, all of them hurrying to get beyond the dangerous pass before the first snows fell. The pioneers of the Donner-Reed group were the stragglers; they brought up the rear of the westward migration for that year.

After struggling through Truckee Canyon, they emerged abruptly into Truckee Meadows, a garden of grass and good water, a deliverance after unending hardship. But standing high and silent beyond the valley were the Sierra Nevada. The mountains were already white with snow.

Despite the urgent need to push on across the Sierra barrier before the snow deepened, the emigrants rested and let the exhausted oxen graze for several days. Neither the people nor the animals could take much more. After a few days' anxious rest, the emigrants pushed on up toward the pass. The going would have been difficult in good weather with fresh teams, but new snow was covering the grass and the oxen soon grew hungry. On October 31, the first of the wagons were worked around the lake that was later to carry the name of this star-crossed group. A short distance beyond, they first faced the pass. What they saw overwhelmed them with despair. Rising a thousand feet above them was a jumbled wall of granite that blocked the valley like a dam. It lay five feet deep in snow.

They must have been near panic. Food was desperately low and more snow was falling on the pass. They attempted to cross with the oxen and horses, but the deepening snow made it impossible. They turned back and hastily constructed rude cabins and prepared to stay the winter, hoping that by slaughtering the animals, and perhaps with some luck hunting game, they might survive the winter. Thinking of the climate back in the Midwest, they hoped that the snow would melt sufficiently between storms to allow an escape over the pass, or at least enough to make hunting possible.

The snow fell deep that winter—twelve, fifteen feet and more. Even had the emigrants been able to get any distance from the buried cabins to hunt, all game had left the frozen mountains. Once the livestock had been eaten, many people died the slow death of starvation. Others clung to life eating the flesh of the dead. When the winter finally ended, and all survivors had been taken over the pass to the Sacramento Valley, of eighty-two who had encamped near the pass, thirty-five—almost half—had died, leaving the name Donner indelibly written on the granite of the pass.

Even if the Donner-Reed company of emigrants hadn't been trapped by an unusually early winter, they would have had a difficult time scaling the sharp walls of the pass with oxen and wagons; it was hard enough even for pack animals. The pass was really just a ragged notch in the ridge of the high mountains, with a monstrous, steep jumble of rock on the east side.

The pass had first been crossed with wagons in 1844 by a small party of emigrants led by Elisha Stevens. This crossing marks the opening of the California Trail. When the party arrived at the granite wall that stood in the pass, they camped near the lake while scouts searched ahead for a practicable passage. What they found was an exceedingly steep series of rock faces, and boulder fields, and a rushing stream full of jumbled rock. After several days of anxious looking, they discovered a narrow gap in a critical place—a gap wide enough for only one ox to pass at a time. First they carried the contents of the wagons to the top of the pass. They then double-teamed the wagons to the small gap, unhooked the oxen, and led them through the rift. Chains from the oxen above were let down to the wagons, and with great effort the wagons were hoisted, one by one, up the ledge, recoupled, and hauled over the pass.

The exact path where the wagons first crossed Donner Pass will likely never be known. George R. Stewart, historian of the pass, spent probably as much time searching for the old path as the Stevens party had originally and was never able to discover the route. He contended that it was disturbed during the construction of the railroad or the several roads that have at one time or another crossed Donner Pass.

Travelers on the 1915 Lincoln Highway marveled at how easy it was to cross the pass compared with the travails of wagon pioneers, but they had a few challenges of their own. An improved road of sorts existed, but it was steep and rough. It passed a short distance from the site of the emigrant tragedy, skirted the shore of Donner Lake, and climbed the face of the pass. It was for the most part the remainder of the old Dutch Flat and Donner Lake toll road, with many loops and turns. At a dizzying height above the lake to the east, just before crossing the pass, this road crossed the rails of the Southern Pacific. The gravel and dirt road climbed and curved sharply approaching the crossing, and on one side was a great drop to the rocks below. To make matters worse, the tracks disappeared into snowsheds a short distance on either side of the crossing, making it impossible to see trains any distance away. Prudence suggested that motorists stop before crossing the tracks to send someone ahead on foot to peer into the darkness of the snowsheds for approaching headlights, but once stopped on the steep grade, any heavily laden auto would have difficulty starting again, and a backward roll of even a few feet would mean disaster over the edge.

Thornton Round and his family had a difficult time of it here in 1914. They were eastbound and therefore downgrade when they came to the crossing. Round shut off the engines of both of their cars and listened for trains. He held a pencil to the rail to detect any vibration. The way was clear, so they started across first with the low-slung Winton, which immediately became hung up because of the drop-off on the far side. "Needless to say, a railroad track is no place to tarry, and we were scared stiff. We asked Mother to take the baby and leave the car immediately. Then we tried bouncing the car to release it from its perch. This helped a little but not enough." Finally, by dint of frantic shovel work and vigorous rocking, the car was freed before

any trains approached. The other car, a Model T Ford, then followed, crossing the rails with ease.

Though Donner Pass has long been a place of difficulty for travelers, it was recognized early as the best of several difficult options. Several other passes have carried wagon pioneers, gold seekers, and stagecoaches to California, and for a time during the gold rush, Donner lay virtually unused. Then in 1859, Theodore D. Judah, an engineer and supporter of the transcontinental railway, determined that Donner offered the least difficult route for a railroad across the Sierras. By 1863, construction of the Central Pacific had started east from Sacramento, as the Union Pacific began laying track west from Omaha. Because construction in the rugged Sierras was going more slowly than anticipated, and because of a desire to spur trade and settlement along the path of the railroad, a wagon road was constructed from the foothill town of Dutch Flat eastward over the pass to Donner Lake. This road saw heavy stage and freighter traffic until the rails of the Central Pacific were completed over the pass in 1868, and then it fell quickly into disuse.

For some forty years Donner Pass echoed only to the pulse of locomotives climbing the grades, and the Dutch Flat/Donner Lake wagon road fell to weeds and erosion until the automobile made its first ventures into the Sierras. The route of the early Lincoln Highway closely followed the old wagon road over the pass and down the slope toward Sacramento. By 1926 a two-lane engineered highway had been built over Donner, again leaving the old wagon road to disuse. This was U.S. 40, the main auto route to northern California until Interstate 80 was constructed two miles north.

The descent from 7,135-foot Donner Pass to Sacramento at a mere fifty feet above sea level was accomplished on the long, gradual slope of the west face of the Sierras. Like an enormous wedge, the mountains rose sharply from the east, presenting the rock wall that so daunted wagon pioneers but that, west of Donner, slanted gradually into the Sacramento Valley.

After unending miles of sun in the shadeless desert, Lincoln Highway travelers found themselves amid the cool, pungent forest of the moist west slope of the Sierras. They drove beneath tall pines made all the more stunning by previous long days when they had seen few or no trees at all. Alice Ramsey crossed via the Pioneer Branch, but her impressions hold for the main road as well:

Victory was in sight. We had passed the worst of our road problems, and the heat too. All around us were mountain peaks and each time we stopped the views became more extensive and more gorgeous. Trees were plentiful and greener than we had seen for a long time. It was incredible that so soon after leaving that arid, barren desert we could be refreshed in the cool shade of towering evergreens. That was balm to our bodies and spirits. Even the inarticulate Maxwell appeared to echo our sensations.

Majestic sugar pines, Douglas firs and redwoods lined our roads on both sides. What a land! What mountains! What blue skies and clear, sparkling water! Our hearts leapt within us. None of us had ever seen the like—and we loved it. We almost chirped as we exclaimed over the grandeur that surrounded us on all sides. We started talking over plans when the trip was completed.

That last great summit marked a watershed of attitude as clearly as it did a watershed of geography. West of Donner Pass, transcontinental auto travelers knew that the rest of the trip would be an easy downhill ride. There were no more high passes to cross, no more vast, uninhabited country, and no more dirt, sand, and mud for roadway. In fact, by 1916, the road west of Donner Pass was well maintained to the standards of the day, and a good portion was even concrete, the first rural paving Lincoln Highway travelers had seen since Illinois.

The realization that the end of the trip was at hand loomed large in the minds of travelers as they dropped toward the huge interior valley of California; most could barely wait to see the vast Pacific, though it was still 127 miles beyond Sacramento. They no doubt thought less about how worn the tires were or about any new sounds the car might make, knowing that help was near and the end was in sight. They also reviewed a few of the high and low spots along the road behind them and thought about the stories they'd tell family and friends back home. For some, the attainment of the Golden State meant the half-way point in a round trip, but for many in the early days, the car and the travelers would return home on the train.

The downhill ride to Sacramento passed through several climatic and floral zones: from near arctic to Mediterranean, from dense, cool forest to parched and hot farmland in the valley near Sacramento. The first palm trees appeared near Auburn. In Sacramento, travelers passed near the spot where, in January of 1863, Governor Leland Stanford of California had turned the first spadeful of earth for the transcontinental railroad.

Sacramento to San Francisco—127 Miles

Via Galt, Stockton, French Camp, Tracy, Altamont, Livermore, and Oakland. The road proceeds south through the great interior valley of California for sixty-four miles, then turns west near Banta. The road crosses the low Diablo Range at Altamont, elevation 737 feet, and proceeds through the hills to the bay landing in Oakland. It goes by ferry to the foot of Market Street, San Francisco, and via city streets to Lincoln Park.

The Lincoln Highway made a great jog south for many years and actually came into Oakland from the southwest. The excursion from a direct path between Sacramento and San Francisco was necessary because of the many arms and channels of San Francisco Bay. A direct path was opened in 1927 with the completion of the Carquinez Bridge at Vallejo, and the Lincoln Highway soon moved to the more direct route through Davis and Vallejo, saving forty miles.

Before the bridge was built, travelers drove south through the dry, yellow, and brown San Joaquin valley, which was unpleasantly hot and flat after the Sierras. After they turned west again near Banta, the climate began to change as the cool coastal breezes began to filter over the hills of Diablo Range—one of the Coast Ranges. The last stretch of mountainous road began beyond Tracy, but the climb was not great and the road was well improved. At the tiny town of Altamont, the highway crossed the last summit on the way to the Pacific. Soon the Lincoln Highway was absorbed by residential streets in Hayward and Oakland, and finally descended to the ferry landing in Oakland. Most of the last several miles of the Lincoln Highway were traveled via the car ferry to San Francisco.

Today this route is primarily a collection of unnumbered local roads. Much of it has been absorbed to become suburban streets where all trace of the highway is lost.

The official end point for the Lincoln Highway was in Lincoln Park, another six miles west of the ferry landing at the foot of Market Street. Here, at the northwesternmost point of San Francisco, high above the surf of the Pacific, the Lincoln Highway ended opposite the Palace of the Legion of Honor at a small monument marking the spot.

Like the official beginning at the opposite end of the continent, this final punctuation mark for the highway was lost in a great city, lost amid architecture, street, and cityscape. Most residents probably knew that the highway ended in their city, but few could find the exact spot. In the early days of the American highway, before numbers, before widening took the front yard, before limited-access expressways, a highway in a city was simply a street that had more than one name and usually a few more signs than average. Here the Lincoln Highway gave up its identity: San Franciscans knew the last few miles as California Street.

III
Roads for the Country

7
America Takes to the Highway

A journey from the Atlantic to the Pacific by motor car is still something of a sporting proposition. It differs from a tour of the Berkshire Hills or any of the popular, extensive drives in the northeastern part of the United States. Any such accommodations and roads as the eastern tourist is accustomed to must not be expected. You must cheerfully put up with some unpleasantness, as you would on a shooting trip into the Maine woods, for example. Yet there are no hardships nor experiences which make the trip one of undue severity, even to a woman.

—Complete Official Road Guide
of the Lincoln Highway, 1916

Though the Lincoln Highway had been in no sense complete in time for the 1915 Panama-Pacific Exposition as Carl Fisher had dreamed, thousands had driven the road after it dried in June. Harry Ostermann, field secretary for the Lincoln Highway Association, once estimated that prior to 1912 there had been fewer than a dozen coast-to-coast auto trips. Probably an underestimation, but even doubled or tripled that figure is small compared to the estimated five or ten thousand autos that drove the Lincoln Highway to the fair in 1915.

During those early years up until the First World War, the association gained strength, and the name Lincoln Highway became a household word. In 1915 the association produced and sold the first *Complete Official Road Guide of the Lincoln Highway,* and followed it with the second in 1916. The association continued to sell buttons, pennants, framed copies of the proclamation, radiator emblems, maps, and paperweights to motorists, businessmen, schoolkids, and patriots.

By 1916, local communities, with their own donated capital and some county and state money, were working to make what improvements they could on the highway. They installed culverts, reinforced bridges, and built up some embankments. After rains, when the mud began to dry, farmers pulled drags back and forth to smooth the ruts and speed evaporation. With the removal of a few kinks in the route, and the realignment in Ohio, the route was shortened by fifty-eight miles, from the original 3,389 to 3,331.

The road was also a good deal easier to follow. Under the direction and sponsorship of the Lincoln Highway Association, caravans of painters had crossed and recrossed the country with brushes and stencils and had painted telephone poles with the Lincoln Highway emblem—red, white, and blue stripes with a large letter L. These official markings stood along the road with the Lincoln Highway Garages, Lincoln Highway Hotels, Lincoln Highway Cafes, and the streets renamed Lincoln Way, making the way of the traveler increasingly obvious.

*A Lincoln Highway staff meeting somewhere
on the open plains of Wyoming, 1916. Field
Secretary Harry Ostermann (third from
left) confers with the official sign-posting
crew.*
Courtesy of the University of Michigan.

With the influx of auto travelers, prosperity began to return to tired
mining towns, according to a bulletin sent to the association directors,
founders, and officers in 1916:

Small towns to the west of us which have been described in the Saturday
Evening Post *as "ghost cities of the west" are coming to life, especially
during the summer months and hundreds of people who have barely lived
for years, are losing their abdominal wrinkles and discarding the old pipes
for real cigars.*

*A Hotel in Eureka, built of stone with running hot and cold water in
every room, was closed for years until the advent of the auto tourists over
the Lincoln Highway converted it again into a profitable institution.*

The Lincoln was popular not only with auto travelers and residents
of towns along the route, but with schoolkids, hobos, and the Grand
Army of the Republic. In 1916, the Women's Relief Corps, the auxiliary of the G.A.R., planned a campaign to provide every schoolhouse
along the Lincoln Highway with an American flag of standard size.
Not to be outdone, a group of Daughters of the American Revolution
proudly announced their intention to plant floral flags in city parks
and on lands both public and private clear across the land. Since
everything stationary and more than a few things on wheels had been
painted up to become red, white, and blue Lincoln Highway markers,
communities turned to tree plantings and street renamings to express
their patriotic pride in the highway. At about the same time DeMers,
"the Hobo Magician," walked the entire length of the highway pushing a handcart. A pair of women rode motorcycles from New York to
San Francisco to prove their fitness as dispatch bearers in case the
U.S. should enter the war. And the Lincoln Highway Association announced that honeymooners were flocking to the highway as if it
were a new Niagara Falls.

On July 11, 1916, affairs relating to good roads took a decided turn for the better when President Woodrow Wilson signed into law the Federal Aid Road Act of 1916, the first of many that eventually would see the highways of America built at public expense. This act was the first to contain any real funding for the nation's roads as a whole. It appropriated some $75 million to be spent over five years to improve rural post roads. The money was to be spent under the jurisdiction of the state highway departments, and since six states had none, the act precipitated a scramble in those capitals. The funds were to be granted on a matching basis, paying up to 50 percent of a project.

While this landmark act for the first time provided real capital for highway construction and improvement, it did not require that it be spent in any sort of organized fashion on important routes; the money could be diluted and spread over a state's entire mileage, yielding very little actual improvement. Nor did the act require that there be any coordination from state to state as to which routes would be improved. Under this new bill, the improved road of one state could end at the state line in a mudhole or at a fence. But it was a beginning. It demonstrated the first commitment on the part of the federal government to get America out of the mud. The directors of the Lincoln Highway Association nurtured a growing hope of seeing their road completed.

Meanwhile, the association boasted about such an increase in traffic on the Lincoln Highway through Princeton, New Jersey, that road officials there were looking for ways to divert it. Heavy traffic would soon cease to be the stuff of boasts and become a problem more vexing than merely getting paved roads for America.

Before much of the money granted by the 1916 act could be applied to road building, the United States was swept into World War I, and most road projects stopped. Any work beyond what was necessary to keep the roads passable ground to a halt as men set aside shovels for rifles. Auto touring slowed considerably as all efforts turned to defeating the Axis powers. Though the promotional work of the Lincoln

Highway Association continued through the war years, little actual highway improvement was accomplished.

In 1918, while the war raged in Europe, Frank A. Seiberling succeeded Joy as president of the Lincoln Highway Association when Joy joined the army as a member of the wartime industrial staff. Seiberling had been vice-president and director of the Lincoln Highway Association and had stayed at the middle of Lincoln Highway activities ever since he made his enthusiastic pledge of 300,000 Goodyear dollars after Fisher's kick-off speech in 1912.

Seiberling was a good choice to head the association. Like Joy, he was in the mainstream of the automobile industry and was a well-respected company president. He had founded Goodyear Tire and Rubber in 1898 with $3,500 of borrowed money and an abandoned strawboard factory. By 1918, he had built Goodyear into the largest tire manufacturer in the world. While his financial acumen set the company on an even course, it was his inventive genius that caused its rapid growth. Seiberling coinvented an important tire-building machine, developed the straight-sided and cord tires, and perfected the detachable rim to the heartfelt thanks of every auto owner who ever changed a tire.

Frank Seiberling's greatest concern was the need for and the difficulty of building roads in the unpopulated West. Though not an enthusiastic traveler in the same spirit as Henry Joy, Seiberling had been to the West and had seen how much work actually needed to be accomplished in the desert and mountains.

By the time the United States entered World War I the Lincoln Highway was popular and usually passable, but it was certainly no paved boulevard. East of Chicago and in California, a few tendrils of hard surfacing were finding their way from town to town, generally built out of state and county road funds and aided to some extent by the Lincoln Highway Association. But in Iowa and the interior West, the road was little better than it had been in 1913 when Joy criss-

Frank A. Seiberling.
Courtesy of Goodyear.

crossed the country to pick the route. Iowa's section of the highway was still dirt, though some paving had begun in Nebraska and Illinois. In Iowa the problem was a legal structure that assessed adjacent landowners a great portion of road construction costs. As long as the law stood, and farmers voted, no road improvement bill would be approved in Iowa.

Out West the problem was different. In thinly populated states like Nevada and Wyoming the distances were great and there simply wasn't sufficient tax base to improve so many miles of road. The terrain was often inhospitable, and great cloudbursts made short work of any road improvements. But in western Utah, the problem wasn't a matter of road improvement: west of Salt Lake City, there was hardly any road at all.

The road between Salt Lake City and Ely was the least improved and least direct section along the entire highway. Because of the terrain and lack of habitation, Henry Joy had been forced to take the existing road—the historic but winding pony express and overland stage trail. The ink was barely dry on the first association maps when efforts began to straighten the road, and by 1916 the directors had figured out a way to do it. The plan required construction of an eighteen-mile-long embankment straight across a southern lobe of the salt flats between Granite Mountain and a point east of Gold Hill, plus six miles of new road in Johnson Pass in the Stansbury Mountains. This work would shorten the Lincoln Highway by a substantial forty-eight miles and eliminate the most crooked and rough sections through Fish Springs and Callao. It would also eliminate Thomas's mudhole.

Late in 1916, Seiberling offered $100,000 from the coffers of Goodyear, and Carl Fisher offered $25,000. The money would go to the state of Utah, and its road commission would do the work. An elevated and graveled grade was the intent here; paving would have to wait for federal money. Here was real money for real road improvement that could shorten the Lincoln Highway dramatically and eliminate some of the worst road.

But the state of Utah balked. It wouldn't accept the money. Governor William Spry had other ideas that didn't include the Lincoln Highway, and in fact didn't include any road running directly west from Salt Lake City. What Spry and his road commissioners pushed for was improvement of a highway that angled south-southwest from Salt Lake City toward Los Angeles—they weren't much concerned with the wishes of San Francisco–bound traffic. This road—variously called the Arrowhead Trail and Pikes Peak Ocean to Ocean Highway—kept the traveler within the state of Utah for several hundred more miles and several more days than did any road that ran due west from Salt Lake City. This path would entice the tourist to Zion and Bryce Canyon parks, and would insure that travelers had greater opportunity to spend money in the Beehive State. Via the Lincoln Highway, travelers would pass through only a few scattered ranch communities and soon pass out of the state, taking most of their money with them.

To weaken the position of the Lincoln Highway proponents, Spry encouraged the improvement of the miserable path that went west of Salt Lake City across the salt flats to Wendover. This road did eliminate many of the mountainous stretches that the Lincoln traversed, but it had the distinct disadvantage of crossing forty miles of the worst salt desert. Here, the salt plain was much lower and wider than where the new Lincoln Highway route would cross, and in the spring this entire area usually was submerged for miles around. Furthermore, travelers were on their own: towns, settlements, and drinkable water were unknown for a distance of some fifty to seventy miles. The road did parallel the Western Pacific Railroad, so at least stranded motorists could flag a train.

Part of the route was across salt flat; part was across mud flat. The mud was usually impassable, but ironically, the undisturbed salt made a fairly solid surface even when it was entirely submerged. When the conditions were just right during high water in the spring, locals who knew the route well would sometimes venture across in "seagoing flivvers," and could be as far as twenty miles from dry land. Holes in the salt floor must have caused some great moments, to say nothing of the corrosive powers of that much salt; but most of all, it just wasn't the sort of path that the traveler from Cleveland would willingly accept. What about changing a tire? In its unimproved condition, this simply wasn't a legitimate road.

For this to be a road in any sense, a high and strong embankment would have to be built for the entire distance, a much costlier option than the proposed construction on the Lincoln. Certainly Spry knew that it would be many years before this road would be improved substantially, and to insure that, the state budgeted only $40,000— enough for a few culverts and some embankment construction, but little else. Yet the state of Utah promoted it as the better route, and few people in Salt Lake City bothered to see for themselves.

What further turned Utah against the Lincoln was that, even though it was headed for San Francisco, it provided a better gateway to Los Angeles than did the Arrowhead. From Salt Lake City to Los Angeles it was nearly sixty miles shorter via the Lincoln and Ely than via the Arrowhead and southwest Utah.

The Lincoln Highway Association recognized its difficult position. If the Arrowhead Trail was built to Los Angeles, the Lincoln would lose much of the transcontinental traffic. And if the Wendover route was built, the Lincoln would never be finished. There would be no direct connection to Ely and the Nevada section of the Lincoln Highway; all traffic would flow through Wendover and across Nevada on the rival highway.

The association mounted its case based on the impracticality of the Wendover route, the unacceptability of the Arrowhead Trail as a feasible route to San Francisco, and the fact that the association had the cold cash to make the needed improvements on the Lincoln. When a new governor, Simon Bamberger, was elected in Utah, Seiberling and Nevada consul Gael Hoag moved in to convince him that his predecessor had been wrong.

In the spring of 1918, after considerable effort and negotiation, the state of Utah and the Lincoln Highway Association signed a contract for the construction of the new Lincoln Highway route across the southern lobe of desert, a route by now called the Seiberling Section or the Goodyear Cutoff. The checks were passed and construction began. The contract called for Utah to finish the road by July 1, 1919.

The deadline passed with work still under way, but inasmuch as good progress was being made and the state of Utah had demonstrated a commitment to the job, the association said nothing. Then, in September of 1919, Nevada consul Hoag discovered that all work had stopped and that the equipment was being moved out. The embankment had been completed across the salt flats, but only seven miles of road had been graveled. Seiberling wrote Governor Bamberger and demanded an explanation. After several days, the governor replied that the road machinery was in need of repair. Further inquiry brought the response that the state of Utah lacked money and would complete the work as soon as funds were available. Continued inquiry brought no additional response.

This had become Frank Seiberling's challenge, first as a director of the association, then as president. Seiberling was known in business circles as "Little Napoleon" partly because he was only five-foot-three, but mostly for his undying will to win. Once convinced of the merit of an idea or plan, he would push until the goal was reached. There were still cards to play in the game of Utah roadsmanship.

While the battle over the highway route was going on in Utah, the big war was going on in Europe. World War I, though it slowed work and travel on the highway, provided an opportunity for Seiberling, the Lincoln Highway Association, and Goodyear Rubber.

During the peak of conflict, the railroads of America were jammed with the materiel of war. Railroad cars loaded with trucks, ammunition, supplies, equipment, and troops choked the main lines everywhere and overwhelmed the railyards of the east coast embarkation points, causing long delays in ship loading. In an attempt to relieve pressure, many shipments, including new military trucks bound for ports in New Jersey, were diverted to the roads. Coordinating this highway transportation effort was Roy Chapin, president of the Hudson Motor Car Company and a director of the Lincoln Highway Association. He was appointed chairman of the Highway Transport Committee of the Council for National Defense and oversaw the operation of some eighteen thousand trucks.

State and county road commissions worked hard to keep the roads open summer and winter for trucks hauling supplies for the campaign in Europe. Trucks also moved less-than-carload freight between cities in the East and perishables from farm to city, lifting a burden from the railroads and alleviating food shortages caused by congested rail lines. Chapin's Highway Transport Committee established bureaus to coordinate shipments and to act as clearinghouses for return loads lest the trucks return home empty.

Trucks had been around since the beginnings of the automobile, but before the war most trucks on the roads had been used for intracity delivery. They hauled milk, groceries, ice, lumber, and all manner of goods from one place to another within a community. They had proven themselves to be faster, cleaner, and more economical than horse and wagon, but once they left the city limits, the railroad had the distinct advantage in speed, load size, and reliability. Bad roads and primitive tires kept trucks lightly loaded and crawling at Conestoga wagon speeds.

A light truck was built on an auto chassis and had automobile tires,

*Tama, Iowa. A well-known Lincoln
Highway landmark, this bridge was built in
1915.*

Old U.S. 30 / Lincoln Highway, Marshall County, Iowa.

So on we plunged deeper and deeper into Iowa, and into possession of more and more of her magic acres. The sun was gradually getting ahead of us, and toward evening came coursing down the sky plumb into the Lincoln Highway, hit a stretch of hard surface, and bounded along before our car until it disappeared over a crest against the sky, its tail-light leaving a yellow, lambent flame within our westering eyes.
—Dallas Lore Sharp, 1927
    The Better Country

Grand Junction, Iowa.

Greene County, Iowa.

Scranton, Iowa.

Right hand road; turn right. Cross RR at Scranton 94.4. 4-corners, blacksmith shop on left; turn left. 4-corners; turn right 1 block, turning left with travel at end of road just beyond.
—*Official Automobile Blue Book, 1917*

Honey Creek, Iowa.

There is constant change of scenery that
makes any little inconvenience that might
have been previously encountered appear as
worth too little for the wonderfulness of
these miles of grandeur. From Missouri Valley through Loveland and Honey Creek to
Cres[c]ent and Council Bluffs is indeed a
magnificent motor trip, and where ever one
travels, there will always linger the memory
of these last few miles of Iowa's Lincoln
Highway.
—local Lincoln Highway guide, 1922

Missouri River crossing, Blair, Nebraska. Chicago and North Western Railway and U.S. 30 / Lincoln Highway. The highway moved to this route, bypassing Omaha, with the completion of the Abraham Lincoln Memorial Bridge here in 1930.

Nebraska carries the motor tourist out of the last of the East and into the beginning of the authentic West. Its corrugated roads possess no apparent upward incline, but they lift the traveler imperceptibly from a one thousand to a five thousand foot altitude in the length of the state. Nebraska is the green of earnest cultivation at its Missouri Valley end and fades out into tan and dun—the feline hue of savage, untamed things—at its other. From carefully tended farms it runs at last into the open prairie, a foretaste of the sun-soaked, wind-brightened immensity of brown empty land that is Wyoming.
—Frederic Van de Water, 1927

North Bend, Nebraska.

Rolled into North Bend to the Corner Cafe for breakfast, a wonderful old eatery with a large horseshoe counter. A booth near the window provides a place to read the paper and write a couple postcards. Bacon, eggs, toast, American fries and coffee come to $2.69. Postcards and an Omaha World Herald, *another 40 cents. I take photos of the cafe and then push west.*
—Drake Hokanson, 1982
    unpublished travel journal

1919 brick Lincoln Highway, Elkhorn, Nebraska.

Shelton, Nebraska.

Black. All-steel body. One-man top.
Weather proof side curtains opening with all
four doors. Four cord tires, nickeled head
lamp rims, wind shield wiper. Starter and
demountable rims $85 extra. Balloon tires
$25 extra. Price f.o.b. Detroit. $290.
—ad for Model T touring car, 1926

It was the miracle God had wrought. And it
was patently the sort of thing that could
only happen once. Mechanically uncanny, it
was like nothing that had ever come to the
world before. Flourishing industries rose
and fell with it. As a vehicle, it was hard-
working, commonplace, heroic; and it often
seemed to transmit those qualities to the
persons who rode in it.
—E. B. White, 1936
  "Farewell, My Lovely!"

Kearney, Nebraska.

Just west of Kearney stands a big sign say-ing 17 hundred and some miles to Boston and 17 hundred and the same to San Francisco. Near here we purchased a new shoe, repaired several punctured tubes, and drove on to Cozad where we camped for the night.
—D. Norman Longaker, 1923
   unpublished diary

Ogallala, Nebraska.

California Hill, Brule, Nebraska. Thousands of wagons bound for Oregon and California climbed out of the South Platte valley here in the 1840s and 1850s, leaving a path that can be seen as the taller vegetation on the hillside.

Farm directory sign, Big Springs, Nebraska.

We had just splashed through the mudholes at a place called Big Springs when we ran up a longish grade and came out on the top of a tableland where for the first time we found ourselves on the plains—the western prairies of song and story.

As I looked all around and saw nothing but space—and so much of it—a feeling of loneliness came over me. Without realizing what I was doing I stopped the car and shut off the engine.

I listened intently for the music of the spheres, the whispering of the stars passing overhead in the daytime, the swishing of the planets through the immeasurable nothingness—but I couldn't hear a thing.
—Bellamy Partridge, 1913
Fill'er Up

*Union Pacific Railroad and U.S. 30 /
Lincoln Highway, Cheyenne County,
Nebraska.*

*From grain elevator to grain elevator, we
laid our course through Nebraska, traveling
over a terrain that grew continually
stranger to eastern eyes and a heat that
amazed and almost frightened even New
Yorkers who have spent summers in town. It
smote us the morning we left Grand Island
and remained with us, despite increasing altitude, for thirty-six hours.*
—Frederic Van de Water, 1927

Bushnell, Nebraska.

BUSHNELL

N.Y.   S.F.

1909   1422

Pop. 100. Kimball County. One hotel, 1 garage. Railroad station, 1 express company, 1 telephone company, 1 bank, 12 business places.

—Complete Official Road Guide of the Lincoln Highway, 1916

which were going farther between flats and blowouts by 1918; but heavy trucks, like those of Chapin's Council for National Defense, were limited to solid rubber tires and snail-like speeds. These crude tires were only a little better than bare steel wheels, and at speeds above fifteen miles per hour they either flew apart in jagged chunks or pounded so badly that any load more delicate than lump coal would be hopelessly damaged. At speed, drivers would have great difficulty just hanging onto the steering wheel and staying in the cab. Except for during wartime emergency, then, heavy trucks really couldn't be of much transportation value over long distances; at fifteen miles per hour maximum on the roads of the day, they simply couldn't compete with the railroads.

Under the leadership of Frank Seiberling, Goodyear saw the situation differently. In 1916, Goodyear had produced and marketed the first practical pneumatic tire for heavy trucks, but despite its obvious advantages of allowing higher speeds and providing much smoother

In 1918 Goodyear began transcontinental trips with its fleet of Wingfoot Express trucks. In the noontime summer sun of that year a Goodyear truck makes a meal stop at Orr's Ranch, Utah.
Courtesy of the University of Michigan.

ride, no one wanted it. Truck operators and the military could still see many auto drivers pulled to the side of the road changing tires; why should they throw away their solids, which were fairly reliable, for these new pumped-up tires that were sure to let them down? That year Goodyear sold only one pneumatic for every sixty-three solid truck tires.

The following year Goodyear launched the Wingfoot Express, a truck line intended to haul tires and materials for the company, but more importantly, to demonstrate the usefulness of these fat new tires. Seiberling's hand is apparent here: the first truck was a five-ton Packard emblazoned with large Lincoln Highway emblems. It also carried the Goodyear name and logo and implored the nation to "Ship by Truck." In its specially built van body was the first sleeper cab in the industry. The truck carried no freight on the first trip—a pile of spare tires, a compressor, and the usual rope, shovels, and material necessary for inventing ways out of bad spots filled the back. The first venture was from Akron to Boston, some 740 miles, and even with two drivers the trip took twenty-eight days and twenty-eight tires, including a long stop at Greensburg, Pennsylvania, where the men waited for a new engine.

By the next year, 1918, many of the difficulties had been ironed out; the tires were improved and several additional trucks built. Using the Lincoln Highway, the Wingfoot Express trucks hauled shoe soles to New England and tire fabric back to the Akron factory; they hauled tires to Chicago and Red Cross supplies from there to shipping points on the East Coast. Through cities, towns, and road conditions of all sorts, they averaged fifteen miles per hour—very good time for the day. They also took Boy Scouts on a three-thousand-mile excursion along the East Coast and made four grueling round-trip transcontinental runs. These trips were difficult, but with each success, Goodyear garnered considerable publicity and learned a little more about building tires to carry heavy loads. And

whenever they could, these trucks—many of them Packards with Goodyear tires—followed the Lincoln Highway of Henry Joy and Frank Seiberling.

The percentage of wartime freight diverted to the roads via the Wingfoot Express and Chapin's Highway Transport Committee was actually quite small. Except for the cross-country trips by the Goodyear trucks, most hauls were fairly short and were concentrated over the more improved routes between cities in the eastern third of the country. The effort did, however, buy the railroads a bit of relief at the time they needed it most. But more importantly, the war and the transportation relief efforts provided the cradle for long-distance trucking. A system of freight transportation had begun; heavy trucks would soon share the road with the automobile.

Meanwhile, the seedling mile program continued during the war but at a greatly reduced rate. Materials and manpower were in short supply, but several sample miles of concrete were built. A seedling mile in Linn County, Iowa, was completed in 1918 and was among the first miles of concrete in the state. In dry weather it became popular with auto drivers who liked to feel their machines hit stride on the fast pavement. But in wet weather, people rode the trolley out to see it; you couldn't drive to it for the mud at either end.

When the program ended in 1919, exemplary miles had been built in Ohio, Indiana, Illinois, Iowa, and Nebraska. During the six years of the seedling mile program, nearly one million dollars in Lincoln Highway Association cash and materials had been spent sowing these seeds of good roads.

What ended the seedling mile program was the realization that the public no longer needed to be convinced about the value of good roads. Seiberling and the directors knew that with every automobile sold, another proponent of paved highways came to be. The automobile had caught on, everybody wanted one and almost everybody who bought one soon got into trouble somewhere. Nothing converted skeptics faster than spending a few hours stuck in the mud far from home.

As the great war in Europe ended, the Lincoln Highway Association under Frank Seiberling dropped the seedling mile program and turned its full attention to aiding construction in the West. While the battle over the Utah route still smoldered, Seiberling returned to Fisher's original list of subscribers, added a few of his own, and set out to find large sums of money for permanent highway improvement in the road-poor West. He was successful: Willys-Overland contributed $20,000 for road building in Wyoming; General Motors bought some $80,000 worth of work in Nevada. In all, some $218,000 of donated money was spent on road projects in the West by the mid-1920s. But this was really only a small portion of what was needed. The directors of the Lincoln Highway Association knew that they could raise enough money to improve some portions of the road in the West, but that only the federal government could build a real highway between Cheyenne and the California border, some 1,100 miles.

With the 1916 highway act, the political winds affecting federal funding for large-scale road construction had shifted. This bill was a great change in policy for the government, which had spent virtually nothing on road building during the past seventy years. Now that the door had been opened, citizens, businessmen, and politicians all clamored for additional federal road funding and for designation of a national network of highways, including the Lincoln Highway. The directors of the Lincoln Highway Association could see the day coming when they would be out of a job, but it remained to them and to the army to stage one last act that would help set the federal machine to building roads everywhere.

On July 7, 1919, eight months after the armistice, Harry C. Ostermann, field secretary of the Lincoln Highway Association, started his white Packard touring car, complete with Lincoln Highway radiator emblem and signs, and drove slowly through a crowd gathered near the White House. Behind him followed a convoy of fifty-six military vehicles, numerous civilian vehicles, 209 officers and enlisted men, and dozens of civilians, bringing the total to seventy-two vehicles and 297 men. After endless speeches and rounds of applause, the convoy departed the Zero Milestone and headed northwest toward the Lincoln Highway and toward San Francisco.

The first Army Transcontinental Motor Convoy was launched as a reliability test of different army vehicles and a training mission for officers and men. It was also an experiment. Nobody really knew if the men could get such heavy vehicles through to the West Coast. It turned out to be a thorough test.

This must have been a high point in life for Harry Ostermann. It was at his urging that the army was undertaking this historic trip. During the war he had piloted numerous convoys of trucks to Atlantic ports via the eastern part of the Lincoln Highway under Chapin's Highway Transport Committee. With the encouragement of the Lincoln Highway Association, and buoyed by success moving materiel by road during the war, the army decided to test the concept fully.

Though Colonel Charles W. McClure commanded the army convoy, Ostermann drove in front because he knew the way. Ostermann knew the Lincoln Highway better than anyone else; he drove it coast to coast as many as three times a year and had seen it more times than even Henry Joy. This man in the shining white Twin-Six was the well-dressed, dignified representative of the association and a sort of folk hero of motoring. In front of small groups or large, before farmers or executives, Ostermann had a persuasive power that he put to work to great benefit for good roads and the Lincoln Highway. His tact and geniality won him thousands of friends across the country. He had begun his professional life very early as a newsboy in New York City, and by the age of nineteen was traveling with Buffalo Bill's

Wild West Show. On this day, the show was different, but Ostermann was clearly the master of ceremonies.

The convoy stretched out behind in an impressive two-mile array of vehicles: army trucks of several manufacturers and all sizes, ambulances, motorcycles, an officers' work truck, a trailer carrying a huge searchlight and another hauling a small example of the new machine that helped win the war: the armored tank. The civilian contingent drove gleaming cars emblazoned with banners and flags, carrying representatives of auto and truck manufacturers and tire makers, representatives of the press and the Lincoln Highway Association. Through the efforts of Frank Seiberling, the Goodyear Tire and Rubber Company's fifteen-piece band accompanied the convoy in one of the company's trucks.

A last-minute volunteer for the transcontinental trip was a young lieutenant colonel named Dwight D. Eisenhower. He joined the convoy just as it was about to leave Washington. He felt lucky to have been selected to make the trip; even luckier, he thought, was to arrive after the speeches and ceremony had ended. Good luck didn't last, however. Eisenhower wrote:

Delays, sadly, were to be the order of the day. The convoy had been literally thrown together and there was little discernable control. All drivers had claimed lengthy experience in driving trucks; some of them, it turned out, had never handled anything more advanced than a model T. Most colored the air with expression in starting and stopping that indicated a longer association with teams of horses than with internal combustion engines.

In the East, the roads weren't much problem, but the trucks acted up like skittish horses. In the first few days, the convoy was held up by broken fanbelts, reluctant magnetos, broken steering parts, sticky valves, but no punctured tires—the trucks rode on solid rubber.

They averaged less than six miles per hour for the first three days.

Despite the slow pace, this was a grand opportunity for America to welcome her dusty doughboys home, and to do it in person in big and little towns scattered across the country along the Lincoln Highway. In every town there were cheering crowds, dignitaries, flags, and bunting. One soldier remarked that the convoy traveled across the continent between two rows of cameras. The soldiers would get as tired of speeches and banquets as they did of the constant clouds of dust and a trail of collapsed bridges. The celebrations got bigger and bigger as the convoy struggled west, to the point that they delayed the entourage nearly as much as troubles on the road.

The festival at Cedar Rapids, Iowa, serves as an example. The weary troops arrived about 5 P.M. after a routine day's trip of minor but ever-delaying and annoying mechanical troubles. After a parade through downtown, they washed, lubricated, repaired, and inspected the vehicles, then settled into an outdoor banquet prepared by a local businessman. They no doubt dozed through the promotional film about the highway and the endless speeches that followed. The city sponsored a dance that night, and every man from town was asked to bring two girls so the soldiers would have someone to dance with. As a nightcap, the troops demonstrated the giant searchlight and, with the assistance of a local airline, staged a mock attack to prove the value of the light in wartime. At daybreak the next morning, the doughboys were up again and getting ready for another day's run.

The roads grew worse as they pushed westward. The dust grew thicker and the country drier, but the drivers and mechanics were learning about what it took to keep the trucks moving. One of the greatest difficulties was the poor condition of many bridges along the route. They were usually strong enough for the ordinary motorcar, but the great weight of these trucks strained and often broke them. The soldiers grew experienced at determining what bridges would hold the heaviest truck and expert at retrieving vehicles from streams when they figured wrongly. During their sixty-two-day, 3,310-mile trip, they fell through and rebuilt no fewer than one hundred bridges. Ostermann and the others from the Lincoln Highway Association must have been pleased at the thought of all the new bridges built at army expense.

The convoy arrived at Lincoln Park in San Francisco on September 7, 1919, to speeches, cheers, fireworks, and banquets.

No doubt Ostermann and the others from the Lincoln Highway Association cheered the loudest. This army trip was just the sort of promotion that the Lincoln Highway needed most. While it demonstrated that the highway was the best route, and legitimized it as the most important route, it also made clear that it was a road in need of much improvement. The army convoy trip had proved that good highways were useful to more than the auto tourist and the small-town businessman; added to the list of good-roads promoters would soon be military staff men, and their voices would be heard loud and clear in the halls of Congress.

It was apparent to the military that if war should ever come again, a carefully planned and built system of highways would be needed. The transportation crisis during the war to end all wars—the first motorized war—convinced even the greatest skeptics that the United States needed a network of all-weather roads.

The stagnation of road funding and the lack of improvement during the war years would be more than offset by this new interest in highways on the part of the military. Under the banner of national defense—not tourism, not trucking, not the Lincoln Highway Association—America would get her paved highways.

In post–World War I America, the notion of automobile touring conquered the nation's imagination. Automobile and auto touring periodicals that had flourished before the conflict moved again into high gear. They were full of ads for machines and accessories, equipment for the long-distance traveler, stories of daring journeys to new, ever more remote places, and encouragement about how easy or cheap or fun it was to climb into your own car and go where fancy may take you.

The lure of distant places always had been strong—travelogues of exotic lands and stories of brave explorers graced the reading tables of most parlors—but opportunity for real travel heretofore had been limited to the rich. Working Americans seldom had money to travel except for business or for an out-of-town funeral. Most Americans seldom ventured far from home; many farmers never left the county. But for the working man or woman, the work week was beginning to shrink, and wages had risen such that there was sometimes a little loose change left at the end of the week after the rent was paid up and the groceries bought, and a person could at least imagine taking a trip in the home country, and imagined doing it in his or her own car.

The automobile that allowed the worker, the farmer, the teacher to do it was, of course, Henry Ford's Model T. The little black car put America on wheels and put Americans on the Lincoln Highway. Though Henry Ford had dealt the Lincoln Highway a serious blow when he refused to contribute to Fisher's ten-million-dollar road fund, he actually did the project a world of greater good by helping to put every man—and most women—behind the wheel of a car.

The Model T was as plain as a workboot, as spendthrift as a farm wife with her egg money, as hardworking as a threshing crew, and occasionally, when the weather was damp and affected the ignition timer, as unpredictable as any Missouri mule. It was cheap to buy and cheaper still to operate. When first introduced in 1908 the open touring model sold for $850, a fine bargain, but by 1926 the price had plummeted to $290—electric starter and demountable rims extra, of course. It didn't take the farmer from Indiana or the machinist from Pittsburgh very long to figure out that this car would go farther on a nickel's worth of gas and a dime's worth of parts than anything else on the road. One auto traveler drove coast to coast in one and claimed he spent a total of $1.90 on repairs. Parts were available at thousands of Ford dealers; it was the only car Ford made, and all Model Ts shared the same basic chassis, engine, and drive train.

The machine also didn't require many parts; it came with few to start with, and they seldom broke because many were made of new, tough vanadium steel. It had a planetary transmission, which writer E. B. White once referred to as "half metaphysics, half sheer friction." It used friction bands instead of the more common, hard-to-shift, straight-cut gears of most cars. The synchromesh transmission was years off, and operators of other cars had to become skilled in the thrashing and timing necessary for successful double clutching when shifting gears. The T had no gearshift; the driver stepped on pedals for low and reverse, and simply moved a lever for high gear.

It was built rugged and it was built to repair easily. Ranchers ran hard over open country herding cattle with it, farmers pulled plows, and people shook their heads when they saw how much of anything it could carry. The little car stood high on thirty-inch tires, giving it good ground clearance, a great advantage on poor roads or in creek bottoms and pastures. The sprightly four-cylinder engine pulled passengers along not so much with dignity as bustling, noisy efficiency.

The Model T didn't break down often, but when it did, it lent itself to simple roadside repairs. Henry Ford felt that a man ought to be able to repair his own car. Though he probably hadn't considered it, many women also became experts. The average tinkerer could, with a little instruction from someone experienced in the ways of the Model T, perform all manner of miracles with simple hand tools. Most operators of the little cars became experts of a sort themselves. The ignition timer, according to White "was an extravagantly odd little device, simple in construction, mysterious in function." Its maladies inspired a particularly varied series of remedies.

Some people, when things went wrong, just clenched their teeth and gave the timer a smart crack with a wrench. Other people opened it up and blew on it. There was a school that held that the timer needed large amounts of oil; they fixed it by frequent baptism. And there was a school that was positive it was meant to run dry as a bone; these people were continually taking it off and wiping it. I remember once spitting into a timer; not in anger, but in a spirit of research.

Left sputtering its loping idle at the curb, a Model T didn't inspire the confidence that a twelve-cylinder Packard did, but it certainly did bespeak a car in a fit to go somewhere. Common folk bought them by the thousands, and as they learned their inner workings and gained trust in Ford's Tin Lizzie, they ventured farther and farther from home in an America now at peace.

When Thornton Round and his family went west back before the war in 1914, they drove two cars—a Model T roadster and a Winton. In the class consciousness of the day, Round referred to the Ford as a "car," the Winton as an "automobile." The Winton used well more than twice as much gas as the Ford, even though the latter car carried the greater weight. Appalled at the miserable gas mileage, the family had an improved carburetor installed on the Winton halfway through the trip, but this improved gas mileage only a little. The big car still guzzled gas at the rate of a gallon for every ten miles. The Winton also burned three times as much oil as the Ford, which is surprising, since at one point the Rounds claim they accidentally drove their little roadster seventy miles without oil. A desert sagebrush knocked open the petcock on the oil pan, draining it unbeknownst to the Rounds. Though the car put up a great racket, they drove on to find repairs. That the engine didn't quickly overheat and seize into a mass of useless metal is a miracle. This abused motor was put back in shape in a mere hour by a blacksmith who installed three new connecting rods.

While the gloved operators of the Wintons, Maxwells, and Packards looked down on the "flivvers," it must have been apparent to the Rounds and millions of others that this much-ridiculed car made more of a transportation dollar than anything else on wheels.

The Model T was well established before World War I, but it wasn't until after the armistice that many Americans set out in them for the far corners of the country. By 1918 the car was ten years old. It had proven itself and was appearing on the road in greater and greater numbers, loaded with travel and camping gear, festooned with pennants, and bearing license plates from every state. It had taken the farmer and his family to town, the working man on the errands of town and urban life. Now it took them to Yellowstone, the Grand Canyon, and the Pacific Coast. The drivers of the big, powerful touring cars were soon sharing the road with the common man in a common car, because by 1920, every other car on the road was a Ford.

By 1919 automobile touring had ceased to be the exclusive enterprise of the well-to-do. Shoestring travelers loaded their Fords with all manner of camping gear, luggage, children, and pets. They bought and studied guidebooks and maps, and after reading of road conditions, added ropes, tire chains, tools, and extra water to the sagging load. They chugged out of the cities and towns and from farms across the country to strike out on their own for the coast, or for any of the wonders of the West that were gaining popularity. On the way, they'd stop and see Aunt May in Iowa or a war buddy in Texas.

They traveled cheaply because there wasn't much money after rent and necessities. Besides, the car itself was a big purchase, and to many, a rather extravagant one at that. There would be no fine hotels or elaborate dining rooms for these new tourists; they bought gas and groceries and figured on fixing the car themselves if it broke down.

They set up camp in schoolyards, pastures, and along streams, often without asking permission, and sometimes leaving a mess when they left. They were gypsies, they were the "motor hobos," the farmers with "go fever"—why shouldn't they be able to come and go as they pleased? Partly as an effort to reduce tent squatting, and partly as an effort to draw this new sort of traveler to town to buy groceries and gas, cities across the country soon began to set up free municipal auto campgrounds. At first these were just large, open areas at the edge of town or down by the river, often with some sort of running water and bath facilities. Travelers would pull in at the end of the day, find an open spot on the grass, and erect a portable home, be it an elaborate tent, a tarp that hung over the car, or a simple ground cloth. While the ritual of homemaking away from home was going on, and just about the time that supper was steaming over the campfire or on the campstove, other travelers would stroll by and exchange comments about the dusty roads or ask about where they'd been or how the car was running.

What began here was a new sort of social interaction, traveler to traveler, sometimes helpful, usually entertaining. Frederic Van de Water identified a new sort of creature among the motor gypsies who angled into tourist camps at night:

A Road Liar is almost any motor camper who possesses the normal human amount of egotism, plus the average automobile tourist's yearning to be considered a hard-bitten adventurer. . . . He is immensely sensitive to thrills and practically immune to the truth. . . . Thus, every road he has traveled becomes, in retrospect, an ordeal through which he brought his car safely only by the possession and exercise of abilities such as few men own. By the time he has made camp at night the difficulties of the day have swelled into tasks before which Hercules himself might have quailed. . . . Any audience will do, and if he succeeds in sending a portion thereof to bed terrified at the thought of the morrow, he feels that his heroism has been proved.

Van de Water found such characters in camps from Iowa to California, and after observing that few of their dire predictions came to pass, offered that "any statements concerning perils ahead can be discounted at once at least 75 per cent. . . . We shall refuse to be scared by their tales hereafter. It may be that henceforth we shall feel sufficiently experienced to do a little road lying ourselves."

Often, though, other campers could be helpful, as Beth O'Shea and her companion found:

Economizing meant camping, and we made our first camp near Clinton [Iowa] on a bank of the Mississippi. It was twilight when we came to the big river, and men in shantyboats were net fishing in the brown water tinged with the pink of sunset. . . .

After a while we came to a large field where an old flivver touring car was parked in a grove of cottonwoods, and a man and a woman were sitting in camp chairs before a fire.

"Is this a tourist camp?" I asked, looking around with interest. We had heard about them, but had expected something more elaborate.

"Sure it's a tourist camp," he told us. "Anythin's a tourist camp what has water and no no-trespass signs. All the towns has got 'em now. They figure they give you a place where you can pitch your tent and you'll buy food and stuff at their stores. You gals want I should give you a hand with your tent?"

Both Van de Water and O'Shea traveled in Model T Fords, and both doted on them as they would a favorite old horse. They often talked to their cars, and called them by name. Beth O'Shea and her companion, Kit Crandall, named their Ford "Henrietta." Van de Water's car was named by his wife. Seeing the little car burdened as it was with heavy duffles on both running boards, she deemed it "Issachar," and when met with a puzzled look from her husband, she told him to consult Genesis, where he found Issachar was "a strong ass crouching

down between two burdens." Another traveler of the era, Stephen Longstreet, traveled coast to coast in a Model T called "Emma," named after a long-departed cook who always overheated when the family had company for supper.

By 1919, most hard-bitten auto travelers boasted a good collection of road guides. These guides had been an indispensable aid to the auto traveler since the first motorized buggies ventured beyond the city limits. The earliest appeared about 1901, and within ten or fifteen years guidebook publishing was a sizeable business. Some were pamphlets; some were heavy books, dense with thin pages. A few were issued for a specific road, like the official Lincoln Highway guides; many others were general and covered numerous routes in an entire region of the country. A few were pamphlets issued by tire and accessory makers or by businesses along a particular route.

Unlike modern guides that primarily detail accommodations and attractions en route, these earlier manuals and pamphlets often described exact routes down to the tenth of a mile, the nature of a particular mudhole, or the color of a house where the main road turned. This level of detail was essential before the days of widespread markings and paving, since such detail was not possible on road maps. A diverging mud lane might be a mud lane, or it might be the transcontinental highway. An ordinary map couldn't clear up the matter, but a guidebook could even tell you what color the mud should be. Gulf Oil had invented the free gas-station road map in 1913, but it wasn't until the 1930s that roads were marked and improved such that maps became the universal navigational aide. Rand McNally published the first *Road Atlas* in 1924, and sales grew as routes became easier to follow.

Schemes other than maps and guides had been tried. The Jones Live-Map Meter of 1909 used a speedometer cable from a front wheel to operate a set of clock gears that slowly turned a ten-inch inter-

Automobile tourists in camp near Omaha,
1923.
Courtesy of the University of Michigan.

changeable disk beneath a pointer. Precise road directions were printed on each disk, and as the car traveled and the disk turned, specific commands—"turn left," for example, or "veer right around the barn"—would be indicated by the pointer, providing the driver with the necessary information at just the right place along the road. Disks were available for many routes in the East and cost ten cents each. Combining parts of different routes meant stopping the car and changing and realigning disks, something that was trouble enough that the Live-Map Meter soon disappeared.

The 1915 *Complete Official Road Guide of the Lincoln Highway* was the first real guide to the highway and had sold well to auto tourists on the way to the Panama-Pacific Exposition and to armchair travelers. The guide had been assembled in a rather hasty fashion, and errors were common.

The 1916 edition was more accurate, comprehensive, and useful. It contained a disclaimer of sorts about the 1915 guide, a foldout map, mileages between towns, descriptions of likely road conditions, data about communities along the route, and ads for hotels and garages. Not surprisingly, there were also ads for Packard, Lehigh Cement, and Prest-O-Lite. The cautious traveler no doubt carefully read the section "Don'ts for Tourists": "Don't allow the car to be without food of some sort at any time west of Salt Lake City. . . . Don't drink alkali water. Serious cramps result. . . . Don't wear new shoes." It also included a three-page list of supplies and equipment the cross-continent tourist should take.

The guide was frank about the mud and lack of improvements in many places, and was careful to point out that while a long tour on the Lincoln didn't require the mettle of Lewis and Clark, it did require careful preparation and an auto in good condition. Although the guide listed mileages between towns, it did not provide detailed, mile-by-mile route information; by 1916 the route was well enough marked that it could, for the most part, easily be followed by noting the red, white, and blue banded telephone poles, rocks, barns, and businesses.

Many of the giveaway pamphlet guides for specific highways—including the Lincoln—were promotional in nature and less objective than the major regional and national guides, or even the official guide to the Lincoln Highway. The traveler sometimes had to be a careful reader to stay out of trouble. "Fourteen miles of good dirt" meant the road was occasionally graded; if the "good" was missing, it meant tough rutted going, and in either case, it meant putting in for the night if it looked at all like rain. One pamphlet guide from 1922 gave this recommendation for the Dining Hall in West Side, Iowa: "An out of the ordinary eating place where good food is prepared and served 'Country Style.' The traveler will find this a real treat. Special Chicken Dinners, Sundays and Tuesdays." Then added disdainfully: "Exterior appearance is the only thing against this place."

But under the seats of most touring cars before the days of numbered highways was a copy of the most respected and useful guidebook: the *Official Automobile Blue Book,* which claimed to be the "Standard Road Guide of America." The first *Blue Book* appeared in 1901 and covered major routes in the East. By 1920, the entire U.S. and much of Canada was covered in thirteen volumes. Each book contained several hundred pages of dense route directions, a foldout map, maps of cities, and ads for garages, hotels, and auto accessories. The route directions were elaborate. The 1917 edition gave these directions for following the Lincoln Highway between Logan and Missouri Valley, Iowa, a distance of just over nine miles:

183.0   0.2   *Logan, 4-corners, at far side of park, bank on left. Turn left 1 block onto 2nd St., turning right onto Main St. at far side of park just beyond.*

183.1   0.1   *3rd St.; turn left.*

183.4   0.3   *End of road, at RR; bear right.*

184.4	1.0	*Left-hand road, signs on left; turn left with one line of poles.*
185.1	0.7	*Left-hand road; turn left with poles.*
186.3	1.2	*Left-hand road; turn left with travel and poles.*
188.6	2.3	*End of road; turn left.*
191.5	2.9	*Fork; bear left.*
192.1	0.6	*Missouri Valley, Erie & 6th Sts.*

It is easy to see why it took two people to drive cross-country in those days. While one person drove, avoiding holes, cows, chickens, and angry farmers with spooked teams, the other sat glued to the odometer and *Blue Book*, shouting distances and directions, continually sorting out left forks and numbers of pole lines, while the car bounced over rough roads and the pages turned in the breeze. An overlooked comma could create a considerable muddle. It's a wonder that any traveler ever noticed the scenery; perhaps only when a wrong turn was made—a "bear right" instead of a right turn—did the driver and navigator stop long enough to notice the surroundings.

Beth O'Shea and her friend Kit tried their trip to the West Coast without a *Blue Book* but realized that more than navigation suffered:

I had refused flatly to take along one of those motor bibles called Blue Books, remembering them too well from the automobile trips of my childhood. "Turn left at the covered bridge," they used to say, "go on two miles to the four forks and bear right by the red barn." Whoever had the seat of honor beside the driver got the Blue Book job and spent the day with his nose glued to the fine-typed pages and read aloud each direction, but never quite in time to prevent the wrong turn. Without a Blue Book we not only took the wrong turns; we frequently kept on wrong roads for miles before we discovered our mistake. That always made Kit furious. How were we going to make Utica that night if we had to spend time doubling back on the trail?

In general, the *Blue Books* were surprisingly accurate, but occasionally things went amiss. Alice Ramsey, crossing northern Ohio in 1909 on her historic trip, was unable to find a certain intersection where she was to make a left turn at a yellow house. Stopping at a farm to inquire, she learned that the intersection was some distance behind. Ramsey was told that the owner of the yellow house was "agin" automobiles and had intentionally repainted it green to disorient motorists.

Meanwhile, as trucks and Model T Fords were being seen on the road in increasing numbers, as people across the land began to think the automobile was something they ought to have for short trips and long, the first generation of the automobile world was aging.

In 1920, Frank Seiberling stepped down as president of the Lincoln Highway Association, and Henry Joy took on the job for the second time. The directors selected Joy for a second stint at a time when the empire of Goodyear was in grave danger of collapse and Seiberling needed every minute to dedicate to his company. His efforts would fail. The company survived, but in 1921 he was forced out of Goodyear, and at age sixty-one started again from scratch with Seiberling Tire.

In the late spring of 1920, Field Secretary Harry Ostermann set out on his twenty-first transcontinental auto trip. As always, he drove the Lincoln Highway Association's gleaming white Packard Twin-Six. With him was his bride of seven months. On June 7, 1920, he and his wife dined with Iowa Lincoln Highway consul D. E. Goodell and friends in Tama. After dinner, Harry departed for Marshalltown, leaving his wife with friends, planning that she would meet him there the next day. He made good time on the graded dirt roads, and near Montour he pulled out to pass a slower Ford. The Packard was doing better than fifty miles per hour when it caught the wet grass at the edge of the highway. The heavy car slid two hundred feet, rolled, and came to rest upright. Before the driver of the Ford could come to his aid, Harry Ostermann, at age forty-three, lay dead on the Lincoln

Highway, crushed by the steering wheel as the soft-topped car rolled over. He was buried in East Liverpool, Ohio, not far from the Lincoln Highway. Henry Joy wrote, "Yes, he's gone on ahead."

Joy served for only six months the second time around as Lincoln Highway president. He was fifty-six and had retired from Packard three years earlier. His health was not good; he could no longer take the strain of business. Though he remained active in the affairs of the Lincoln Highway Association, he otherwise retired to his yacht and experimented with this new thing, radio. Seiberling reluctantly finished Joy's year but could not continue. J. Newton Gunn, vice-president of the United States Rubber Company, was elected president at the next board meeting.

In 1921 the great dream of Carl Fisher was eight years old, and the old guard was changing.

8

Numbers,
Not Names

The traveller may shed tears as he drives the Lincoln Highway or dream dreams as he speeds over the Jefferson Highway, but how can he get a "kick" out of 46 or 55 or 33 or 21?

    —Newspaper editorial,
       ca. 1925

By 1921, a revolution was at hand. The streets, roads, parking lots, and garages of the country held better than nine million cars and trucks. Most of the roads were still unimproved, but it was clear by now that the country wanted paving. And as Americans adjusted to prohibition and to women in the voting booth, they saw Warren G. Harding break with tradition and ride to his inauguration in a motorcar instead of a carriage. The date was March 4, 1921, and the car was a Packard Twin-Six.

In November of that year, Congress passed and President Harding signed the Federal Highway Act of 1921. Like the 1916 act, this bill provided $75 million in federal money to be matched on an equal basis with state funds. But where the earlier act had allowed states to spend construction funds as they saw fit, this bill stated that federal aid should be concentrated upon "such projects as will expedite the completion of an adequate and connected system of highways, interstate in character." This legislation stipulated that each state identify no more than 7 percent of its total mileage as "primary" and that the funds could be used only on these roads. Seven percent amounted to about 200,000 miles of road around the country. Within two years, the skeleton of a national network of highways began to emerge.

This was the sort of legislation that the Lincoln Highway Association had hoped for from the beginning. Though these men directed their efforts toward the completion of a single highway, they saw the Lincoln Highway as an example, a demonstration of a sort of transcontinental road that they imagined would be the first of many. With the Federal Highway Act of 1921, this dream was much closer to being realized. Their work was not done, but now they would have the sort of help only the federal government could provide.

In that year, at about the same time President Harding was adjusting to his new duties in the White House, the Lincoln Highway Association was embarking on a new sort of object-lesson road. While simple concrete paving had been the goal of the educational seedling

*Pouring concrete on the Ideal Section,
October 1922.
Courtesy of the University of Michigan.*

mile program, this new bit of sample highway was to exemplify what was ideal, what the highways of America could and should be. The result was the Ideal Section, built on the Lincoln Highway in Indiana between Schererville and Dyer on the Illinois-Indiana border. It was one and a third miles long with concrete paving laid ten inches thick and forty feet wide to allow four lanes of traffic. Contracts were included in the program for landscaping, lighting, and an adjacent footpath along the 110-foot right-of-way.

The greater part of the $167,000 for construction was provided by the Lincoln Highway Association through a special donation from the United States Rubber Company, a grant arranged by J. Newton Gunn, vice-president of that company and president of the Lincoln Highway Association. The remainder came from state and county funds. When it was completed in December 1922, this short piece of road was hailed as the "finest section of road in the world" and drew the attention of engineers and road builders from across the country.

While it was an ideal for the times, its limitations are all too clear today. In 1922 no one could have foreseen the crush of future traffic. The Ideal Section was narrow for four lanes and was designed for average speeds of only thirty-five miles per hour. The footpath indicates that no one imagined the fumes, noise, congestion, and roadside development to come. Nor did the rustic walking trail winding in and out among the trees account for the fact that in just a few years Americans would go everywhere by car and nowhere on foot.

The section was extensively illuminated for night driving. In a statement that seems quaint and naive today, the Lincoln Highway Association boasted that "motorists were surprised and delighted to find they could drive this section at night, without headlamps on their cars, as fast and as safely as in full daylight, having full view of the road and of approaching cars at all times."

About this time the Lincoln Highway Association found its road paralleled on both sides, from Canada to Mexico, by other announced but generally unimproved transcontinental highways. While the association supported the building of an entire network of roads for the country, it was also vehement that its road be completed first. The association had begun this game of transcontinental highway building and was not going to be upstaged by any Johnny-come-lately. Goodwill gave way to stiff opposition when a competing highway organization sought to gain support or funding at the expense of the Lincoln Way.

By 1922 there was a total of nine named transcontinental highways in some way deserving of the title, including the Lincoln. Across the northern states was the Theodore Roosevelt International Highway between Portland, Maine, and Portland, Oregon, and the Yellowstone Trail connected Boston and Seattle. Transcontinental travelers south of the Lincoln Highway could drive the Bankhead Highway between Washington, D.C., and San Diego; the National Old Trails Road between Baltimore and Los Angeles; and the Old Spanish Trail between St. Augustine, Florida, and San Diego.

Close to and sometimes overlapping the Lincoln Way were three trails—three roads that competed with the Lincoln for travelers and road improvement money. The Pikes Peak Ocean to Ocean Highway ran between New York City and Los Angeles, and shared the same roadway as the Lincoln Highway from New York to Philadelphia. The Midland Trail, which had eagerly stepped into tire tracks left by Carl Fisher's Hoosier Tour of 1913 across Kansas and Colorado, connected Washington, D.C., and Los Angeles. Although the Midland shared routes with the Lincoln between Salt Lake City and Ely, Nevada—the troubled desert section—that organization was nearly moribund and could offer no help in seeking the completion of the joint section.

Lastly and of greatest concern for the directors of the Lincoln Highway Association was the Victory Highway, which like the Lincoln claimed terminals of New York and San Francisco. The men of the association were irked at the Victory, this unwanted bedfellow which claimed to run between the same terminals. But the rivalry between them wasn't because of the shared terminals; it was because of their divergent routes in Utah and Nevada.

The Victory Highway was actually a lesser trail, not quite a transcontinental. It was seldom shown on the maps of eastern states, but in the West it had strong support in Utah, Nevada, and northern California. The thorn for the Lincoln was that the Victory had claimed the rival route west from Salt Lake City across the salt desert to Wendover, and across northern Nevada along the Humboldt River through Elko and Winnemucca and to Reno, in direct competition with the Lincoln. And in what must have seemed like a grand insult, the Victory proudly put up its markers alongside those of the Lincoln across Donner Pass between Reno and Sacramento, claiming the route as its own.

The adoption of the Wendover-Elko-Winnemucca route by the Victory heightened the struggle to see which route would become the federal aid highway across western Utah and across Nevada: the Lincoln Highway or the Wendover road, now the Victory Highway. The Victory Highway would give the Lincoln Highway a run for its money.

None of these transcontinental highway organizations had the sort of flag-waving patriotic following that the Lincoln Highway commanded. Few—including the Victory—had any real possibility of raising substantial money to improve their roads, and certainly none had the sort of money provided by the auto industry that the Lincoln Highway people had already spent. But with federal money on the way, private fund raising became secondary, and even the Lincoln Highway's monetary advantage became less important. Most of the road organizations were promotional and touring organizations which were now working to secure places for their roads in the federal highway system.

Between 1921 and 1925 the Lincoln Highway Association under its third president, J. Newton Gunn, did everything it could to bring a good road to the states of the West. Even though Henry Joy's health

*Traffic was heavy from the start on the
Lincoln Highway in New Jersey, creating
difficulties for construction crews widening*
*the Raritan River bridge at New Brunswick.
Courtesy of the University of Michigan.*

was poor and Frank Seiberling was in the process of starting over in the tire business after being ousted from Goodyear, the two men stayed close to the activities of the association, and in the case of Joy, often acted as mouthpiece for the organization.

By 1922 the new federal money was beginning to build and improve roads, and the association did everything possible to see it directed toward the Lincoln Highway. Certainly the headquarters in Detroit still did a brisk business in guidebooks, memberships, pennants, and radiator emblems, sent to a growing majority of auto owners across the country, but Gunn and the directors focused their attention on the road west of the Mississippi, and in Utah in particular. The tone of the association's bulletins and correspondence was becoming terse and somewhat insistent; there was growing impatience to see this highway completed.

The association fought skirmishes over the routing and delayed improvements in western Pennsylvania and central Ohio. The road from Pittsburgh to the Ohio River was slow, heavily traveled, and needed to be changed to a more direct route. In Ohio, the squabble continued with the Harding Highway people for the route between Mansfield and Delphos.

In Nebraska, the Lincoln Highway Association worked with the Union Pacific to shorten the road across the state by thirty-three miles and to eliminate twenty-four dangerous grade crossings. In the early years, the highway had generally followed the Platte River valley, but on a zigzag course over section roads that often took the traveler around two sides of a right triangle, back and forth across the busy Union Pacific tracks, instead of in a straight line along the river. The railroad followed the terrain and the river closely, and the far preferable course for the highway lay alongside the tracks. Gael S. Hoag, formerly Nevada state consul, now field secretary following the death of Ostermann, used the Union Pacific's own grade-crossing accident statistics to convince the railroad's management that it ought to lease

land along the railroad right-of-way to the state of Nebraska for the highway. Within a few years the road was moved, and except for diversions through towns, the Lincoln Highway came to the shadow of the Union Pacific.

In the early 1920s, the association worked hard with state road officials and community leaders in towns along the road in Nevada and Utah. The Victory Highway to the north through Elko and Winnemucca was growing to be a strong rival—not so much in quality but in promotion—so every possible effort was turned to getting money and labor spent on upgrading the Lincoln Highway across Utah and Nevada. Since the General Motors contribution of nearly $80,000 had fixed the worst sections of the Lincoln—except for the Goodyear section—the effort now had to be to bring the entire road across those states up to an improved standard. Of the two roads the Lincoln was the more improved, but sentiment was turning in favor of the Victory, especially as Utah continued to argue for the Wendover route across the desert.

Meanwhile the Iowa problem continued, and here the words of the association had long been impatient. Back in 1916, Henry Joy, remembering his visit the year before, had lambasted the state for its poor roads: "Today, in the rich state of Iowa, not a wheel turns outside the paved streets of her cities during or for sometime after the frequent heavy rains. Every farm is isolated. Social intercourse ceases. School attendance is impossible. Transportation is at a standstill. Millions of dollars worth of wheeled vehicles become, for the time being, worthless."

By 1922, nothing had changed in the Hawkeye State; it still had only 334 miles of paved road, about 5 percent of the total mileage. Travelers who fretted in advance about the bad roads of Utah still hauled up short and surprised in a three-hundred-mile-long mudhole when the rain came to Iowa. Good-roads groups in the state and across the country, including the Lincoln Highway Association, presented strong and compelling arguments about the cost saving of con-

*Iowa's Lincoln Highway provided
considerable challenge for motorists during
wet weather. In 1919, this mudhole in Story
County was representative of conditions
across the state.
Courtesy of the Iowa Department of
Transportation.*

crete roads. They proved that the cost of construction would soon be returned by the greatly reduced need of maintenance over a dirt road that must be rebuilt fresh after every rain, and they pointed out that tires and vehicles wore out much less quickly on paved roads, but the voters of Iowa stood stubborn. The unequal property-tax structure put a greater burden on the farmer and created a general dislike for the notion of expensive, high-type roads. Even with the 50 percent promise of federal funding, and even when other funding proposals for the state half were offered, the negative rural sentiment carried the day and the roads stayed dirt.

In Utah, the stalemate continued through 1922. Gunn and the association continued to clamor for the completion of the Goodyear Cutoff; the state of Utah continued to do nothing. With the 1921 highway act, the stakes in the Utah battle increased. In most states where the Lincoln went there was little doubt that it would be absorbed as a whole into the federal system. It was the best and most direct road nearly everywhere it went, and local enthusiasm for this highway continued to run high. But in Utah and northern California, the Victory Highway held the popular appeal. If the state of Utah designated the Victory as the federal route, the Lincoln Highway across the Goodyear Cutoff would never be built. Though pressed by the affairs of his new tire company, Seiberling stood at the front of the fight as the association brought all available pressure to bear on the state of Utah and the officials of the Bureau of Public Roads, the administrative agency for federal aid money under the secretary of agriculture, Henry C. Wallace.

Meanwhile, northern California–bound travelers got across the worst part of the salt desert the best they could. In May of 1922, the embankment of the never-completed Goodyear Cutoff on the Lincoln was washed out and impassable for seventeen miles west of Granite Mountain. The Victory Highway to the north was worse: a low, narrow embankment the state had built across the flats for some forty miles the year before was all but gone, and the salt desert was under two feet of water. That year, springtime travelers took the old stage and pony express route—the original Lincoln Highway—through Fish Springs and Callao toward the West Coast. By summer, volunteer labor had repaired the Goodyear Cutoff to driveable condition, and forty cars per day crossed the white desert on the Lincoln Highway's arrow-straight embankment west of Salt Lake City. The only way across on the Wendover route was to drive on the Western Pacific tracks.

Despite the efforts of the association, Utah selected the Victory Highway between Salt Lake and Wendover as its federal highway to the western border. The directors of the association were not surprised, and moved in for the final assault, a last attempt to get the Lincoln designated as the route to receive federal monies. They prepared a book, *A Brief for the Lincoln Highway in Utah and Nevada,* 172 pages of engineering data, maps, illustrations, and argument against the Victory Highway and for the Lincoln Highway as the preferred route between Salt Lake City and Reno. This publishing effort cost nearly four thousand dollars, and Henry Joy footed the bill as a donation to the association.

On May 14, 1923, the final battle for the Lincoln Highway of western Utah was fought in Washington, D.C. Secretary of Agriculture Wallace hosted a day-long hearing between the two factions in hopes of settling the issue. Among nearly one hundred present were Lincoln Highway Association president Newton Gunn, Henry Joy, Frank Seiberling, and Field Secretary Gael Hoag. Making the case for the Victory was the governor of Utah, Charles Maybey, ex-governor William Spry, the Utah state highway engineer, road commissioner, and a former attorney general. People from the states of Nevada and California attended in support of one side or the other.

The Lincoln Highway Association passed out its document on the route and argued for the Lincoln route based on cost of construction,

The arrow-straight Goodyear Cutoff was in
rutted but passable condition in 1921.
Courtesy of the University of Michigan.

time needed for completion, directness, available water, and services.
The Utah representatives argued based on their engineering studies of
the Wendover route.

At issue here was not merely the desert crossing; the outcome
would determine the main federally funded route between Salt Lake
City and Reno. What Utah built would determine which route Nevada would improve. If the Lincoln won out in Utah, then Ely, Eureka, and Austin would see the improved highway come to them. If
the Victory triumphed with the route through Wendover, federal
dollars would build the Nevada highway along the Humboldt River
through Elko and Winnemucca, along the old California Trail.

Despite the claims of both sides, the two routes between Salt Lake
City and Reno were really about equal. Between Salt Lake City and
Reno it was 575 miles via the Lincoln and 569 via the Victory—only
six miles difference. The Lincoln Highway had a distinct advantage in
that it was in much better shape and therefore would cost less to complete. Through the association's efforts and money, the worst places—
excepting the Goodyear Cutoff—had been put in reasonably good
driving condition, and the whole of the Lincoln had seen much more
improvement than the Victory. The *Blue Book* for 1922 indicated that
several parts of the Victory in Nevada were "in poor condition, being
badly cut up with ruts and chuck holes," while the Lincoln was mostly
"natural prairie road" with a few sections of gravel.

The Victory's greatest disadvantage was the forty-mile salt desert
section. Everyone knew that this piece of road would be costly to
complete. So far, the half attempts at building an embankment across
the salt flats had all ended in failure; each year the dense salt water of
springtime was whipped into waves that pounded the berm into sub-
mergence. The *Blue Book* didn't even acknowledge the Wendover desert crossing as a highway.

What the Lincoln had going against it was mountain grades. While
the Victory to the north crossed several summits between Salt Lake

A "sea-going flivver" on the Victory Highway, spring 1922. A Blue Book was insufficient here; an auto tourist might be more comfortable with marine charts and a sextant. Much of the salt desert was crossed on a low and muddy embankment, but for eight miles in the middle, where Utah had as yet made no improvements, the tourist was left to his or her own devices and navigation.
Courtesy of the University of Michigan.

and Reno, the Lincoln crossed five with altitudes greater than seven thousand feet, all higher than those on the Victory, all with deep valleys between, and many with long, radiator-boiling grades connecting them. In at least one place the grade was a dizzying 18 percent. Those grades were taken with ease in a twelve-cylinder Packard, but a Model T or other low-powered car would balk. Hickison Summit on the Lincoln, at 6,564 feet, was more commonly known as Ford's Defeat.

Here is where Henry Joy's route lost favor to the Victory. When he had first come west looking for the Lincoln Highway route, Joy made the best decision possible considering that there was no immediate possibility for a route across the salt to Wendover. Had Joy seen the possibility of improvement money sufficient to complete a road to Wendover, he no doubt would have selected that route and the Lincoln Highway would have followed the Humboldt River through Elko and Winnemucca on the way to Reno.

For the short term, the Lincoln probably was the better road; it could be put into good driving shape with the least amount of money. But by 1923, America was beginning to think about roads and routes for the long haul, roads that would be good for a generation or better. For a permanent system of highways, then, the much higher cost of construction across the salt desert on the Victory would be offset in the long term by avoiding the mountain grades on the Lincoln.

Secretary of Agriculture Wallace offered no decision on the day of the meeting in Washington. The parties and factions went home to await his reply. On June 6, 1923, he announced his choice as the Wendover Cutoff. In a letter to Joy, Secretary Wallace pointed out that under law he had power only to approve or disapprove Utah's choice of the Wendover route; he had no authority to designate any alternative route. He urged the association to adopt the Wendover route as a part of the Lincoln Highway. Joy replied for the association:

We can see neither logic, wisdom nor justice in accepting your suggestion which would now require us to abandon our wise basic principles, to abandon our cooperators in Nevada and Utah, to repudiate our chief supporters, the press and the public.

The Association is not seeking new possibilities for routes. All possible connections we have long ago investigated and the establishment of the present Lincoln Way and our large investments thereon were made only after not only we, but the two states of Utah and Nevada, were fully convinced that it was the only right and proper route and the one which would stand as the ultimate one. The present attitude of the state of Utah does not change the existing facts one iota. Utah's present desire to abandon the Lincoln Way does not make the Lincoln Way incorrect.

But this time Henry Joy was wrong; if a good road could be built across the forty-mile salt desert, the Victory would be the better route in terms of grades, population, and availability of water and supplies.

The Lincoln Highway in western Utah was dead. With new federal money, construction of a permanent earthen berm began between Knolls and Wendover in 1924, and the Lincoln remained a "natural prairie path." The efforts of the association and the money provided by Fisher, Seiberling, and Goodyear were swallowed by the desert sand of Utah, the missing link in the transcontinental highway.

At precisely 1 A.M., the telegrapher gave the signal and Louis B. Miller gunned his engine and pulled away from the curb. A small crowd cheered as the car sped into the night. In the new Wills Sainte Claire roadster—nicknamed "The Gray Goose"—were two men: Miller and his companion C. I. Hansen. They had just left the Western Union office in Jersey City, just two blocks off of the Courtland Ferry to New York. The top was down and the warm summer moon of Tuesday, July 14, 1925, flooded the interior of the car. The two men soon turned westbound into the Lincoln Highway and drove swiftly through the cities of Newark, Elizabeth, and Rahway on deserted streets. They were bound for San Francisco.

Back at the telegraph office, the agent who had given the start signal went inside and keyed a message to C. H. Wills, president of Wills Sainte Claire: "LEFT AT ONE AM JULY FOURTEENTH, L B MILLER." He then went back to more mundane duties. The people who had cheered Miller and Hansen off into the night went home to bed. The two in the roadster couldn't look forward to sleep anytime soon; they wouldn't close their eyes for better than four days, not until they arrived in San Francisco, 3,423 miles away. Miller would drive the entire distance and Hansen would spend his time navigating and keeping the driver awake.

Miller was out to break the transcontinental auto record, and like those who had set and broken it before, he was doing it on the Lincoln Highway. Though other routes existed by 1925, the Lincoln was still the shortest, most direct, and most improved. Because of the publicity that surrounded these record attempts, the Lincoln Highway Association provided maps and road-condition reports, but because most record breakers also broke the speed limits, they felt they couldn't directly sanction the runs. The association also alerted the local consuls along the route in case they could be of assistance.

Since its creation the Lincoln Highway had been the path of almost constant cross-continent speed attempts. The first record of note was set by Bobby Hammond in 1916: an astonishing six days, ten hours, and fifty-nine minutes. Compared with Joy's epic twenty-one-day struggle from Detroit to San Francisco the year before, it is clear what difference a dry road can make. Hammond's record soon fell, and over the years the best time between the coasts dwindled to four days, fourteen hours. This was the record Miller planned to break.

Most of these trips were done as promotions for auto companies—Empire, Marmon, Hudson, Essex, and on this trip, Wills Sainte Claire. While Wills Saint Claire had not sponsored this run, the company certainly had a stake in its success. Small manufacturers like the Wills company were having an increasingly difficult time against the emerging giants of Ford and General Motors, and such a record-breaking trip would get mention in the papers and provide a splendid marketing gimmick. Customers shopped for speed, reliability, and comfort, and delighted in owning cars that held places in the record books. Miller would keep company president C. H. Wills closely advised about their progress with frequent collect telegrams. But for the Wills Sainte Claire company, Miller's heroics were too late: it would be out of business by 1927.

In the predawn silence Miller and Hansen were escorted through Philadelphia by a motorcycle policeman, and at 7:40 they passed Chambersburg, Pennsylvania. The powerful Gray Goose crossed the Alleghenies with ease, passing many cars as it rumbled upgrade. It was noontime when they reached Pittsburgh, and thanks to the assistance of the Rotary Club there, the Gray Goose was led quickly through town, avoiding traffic snarls.

Most of the records for long-distance driving were held by professional drivers—race-car drivers or auto-company men—but Miller was not a racer, nor was he associated with the auto industry in any way; he sold X-ray equipment. Manager for the West Coast operations of the Victor X-Ray Corporation, Miller broke auto records as a hobby. While most men of his station were playing golf, gardening, or learning to sail, Miller was roaring around the West breaking speed records between various cities as preparation for this, his grand safari. He had been warming up for this transcontinental trip for eight years. Miller was fifty-one when he and Hansen made their departure from the East Coast. He would no doubt age more than four days' worth on the way to San Francisco.

The pair quickly crossed Ohio and rolled through Fort Wayne at 10 P.M., stopping only to take on gasoline and cold refreshments. The car was running well; they were maintaining schedule and neither was sleepy. Chicago appeared at daybreak, and with the help of escorts was left behind them before morning traffic grew heavy. By seven they had crossed the Mississippi and entered Iowa at Clinton. The pace was steady; the cadence of the road had become part of them. Twenty-two miles west of there, pavement ended, and except for seedling miles and scattered bits of paving and the brick streets of towns, they'd see no hard surface again until California. It had rained earlier in Iowa, but the roads had dried sufficiently that they were passable. Dry weather held; the Gray Goose crossed the mud state of Iowa in only nine hours, for an average speed of thirty-nine miles per hour. In Missouri Valley, Iowa, Miller scrawled a telegram to Wills that stated, "THE FARTHER THE CAR GOES THE BETTER IT GETS."

Except for a fifty-gallon auxiliary gas tank and a pair of spotlights, the Gray Goose was entirely stock. The car belonged to Miller and was equipped with the then-new Firestone balloon tires which made it possible to negotiate the drying ruts of the Iowa mud at a much higher speed than older narrow tires would have. Owing to lower pressure, they were also much less prone to puncture. Miller and Hansen carried only two spares.

Why the two of them didn't starve to death is a mystery. Some food was passed on board at gas stops, but for the most part they ate oranges out of a bag and ice cream out of thermos bottles, washed down with water. They also carried and drank vast quantities of the stuff that has kept millions of drivers awake before and since: black coffee.

They crossed Nebraska and Wyoming without event, but the strain of endless driving was growing. They passed Salt Lake City and headed west into the desert at nightfall Thursday, almost three days into the trip. The desert was dark; they would see few lights other than their own that night. Fatigue, washouts, and bad roads made progress very slow. Miller had been behind the wheel for a solid three days. Keeping the car on the rough, winding desert road no doubt took great effort, but it must have taken all the concentration of both men to keep each other awake.

When sunrise found them at Gold Hill, Utah, they knew the worst roads were behind them. The Gray Goose began to make better time; Miller and Hansen could smell success beyond the mountains of Nevada. If only they could stay awake. The sheer exhilaration of the trip was beginning to fade; driving at high speed with the top down, the wind blowing through their hair, and the smell of desert and sun could no longer keep them alert.

The fast-changing weather and steep grades of Nevada soon provided some distraction to keep their minds away from sleep. Miller wrote later: "Between Eureka and Austin, just as we started up the last grade, with a seven-mile climb ahead of us, the heavens seemed to open up and the water came down in bucketfuls. We still pounded along, praying that we would reach the top without slipping off the grade, but the good old balloon tires seemed to hang right to the road better than the old style tires equipped with chains."

With night coming on again, they reached western Nevada. To revive their failing energy, they sent word ahead and arranged for a tub of ice water at Carson City, where they stopped briefly for a bracing bath, then pushed on west over the Sierras. A crowd of a thousand people cheered them out of town. They had but 250 miles to go. Miller drove carefully, but lust for sleep was nearly uncontrollable. Though it was cold in the mountains, Hansen put the top down, and as a desperate measure, took to pouring thermoses of ice water down Miller's neck to keep him awake.

At last they reached Sacramento, the mountains behind them, and, hoping to shorten the trip by ferrying across the Straits of Carquinez, they turned off the Lincoln Highway and drove on a diagonal toward Benicia. Their luck ran out at the ferry landing: the rain of Nevada had delayed them enough that they missed the boat and had to back-track sixty-five miles to come into Oakland from the south and east.

This epic trip finally ended when the Wills Sainte Claire approached city hall in Oakland, where a restless crowd had waited all night. The spectators watched the mud-spattered car arrive, complete with a rumpled fender from a tangle with a ditch somewhere back up the road, the only mishap. Miller and Hansen were both tanned, and sur-prisingly, they were clean-shaven. Somehow these speedsters had both shaved each day while rolling cross-country; no doubt Miller's work with a razor in one hand and steering wheel in the other had caused some terror at high speed. Hansen sat slumped in the seat, ready for nothing more than a good meal and a bed, his hair a tangle, looking as though the wits had been scared out of him once too often by the somnolent, fast-driving Miller. Their Spartan road diet must have suited Miller, because when the trip ended he had lost but six pounds, while C. I. Hansen had dropped twenty-four.

L. B. Miller looked tired, but he also looked ready for more. He wrote later: "The trip was over, the old Gray Goose running as sweetly and prettily as it did when we pulled away from the Courtland Ferry in New York, 102 hours, 45 minutes before. For once I had been able to find a piece of mechanical equipment whose endurance would cope with my physical endurance."

He and the Gray Goose still had endurance to test; the record breaker Miller had the gleam of another trip in his eyes as he arrived in Oakland. The record for transcontinental auto travel had fallen, but not by enough to satisfy Miller. He was out to shatter records, not just to break them. The Gray Goose had topped the previous time by only seven hours. If the rain hadn't slowed him in Nevada, causing him to miss the ferry at Benicia, he may have been able to cut an entire day from the time set the year before.

So Miller came back the next year in the same car, now with twenty-three thousand miles on it, and cut almost a day from his

previous time. Hansen, however, was not along. He must have taken up a safer pastime.

Like a farm dog who has tasted blood in the hen house, Miller was never the same again; he couldn't forget the addictive effect of sleep deprivation, speed, and fame. In 1927 he was at it again in a new car, trying a new stunt, trying to break his and anybody else's record. A simple one-way trip was no longer enough; this time he set a record for round-trip driving. Miller drove nonstop from San Francisco to New York City, halted for exactly one minute in New York City—no doubt to use the bathroom—then turned around and rushed back to San Francisco, taking just one minute less than a week.

For America in the decade of the 1920s, Miller, Hansen, and many other long-distance record breakers and racers were the flagpole sit-ters and dance marathoners of the automobile world. They were the devil-may-care fools who set out to prove something, to make a name for themselves, to be remembered for some feat, however zany. The automobile was a natural device for anyone in search of a record; it was still a new thing, so there were many records to be made. Thanks to the Model T, the automobile was democratic enough that nearly everyone could get one. And during the 1920s, almost everybody did. In that decade the motorcar became a part of everyone's world as it wrought more change on the fabric of American life than any single device, idea, or institution before or since.

The map of the United States had, by 1925, bloomed into a dense and unruly matting of named roads. The rising curve of auto sales caused a huge increase in traffic, which made every town want to have a highway; and as soon as a highway was planned, named, or built, it created demand for even more automobiles, which could be counted upon to clog a new road even before the construction bar-ricades came down. There were the major transcontinental routes and the main north-and-south routes, but there were also the minor trails that connected nowhere with somewhere, anyplace with someplace.

The Federal Highway Act of 1921 had brought some order, some coalescence of major routes by requiring that states spend their federal appropriations on only 7 percent of their mileage, but in general, a road map of the United States was still a dense snarl of confusing named roads that all looked to be of nearly equal importance. What had begun with the Lincoln Highway in 1913 had exploded into hundreds of named roads that went from place to place, and like college athletic teams, each had its own color markings. Instead of being emblazoned on jerseys or uniforms, these colors were painted in bands on telephone poles, trees, sides of buildings, or any other stationary object. One transcontinental traveler thought these trailblazings had the appearance of "a far Western brand book," but many autoists were simply confused. The Louisville *Times* griped, "The harmless tourist in his flivver doesn't know whether he is going or coming, whether he is a hundred miles from nowhere or on the right road to a good chicken dinner and a night's lodging."

In the West and at important river crossings these major and minor highways would bunch together, funneling traffic between poles that were brightly banded from the ground to the wires. They looked like totem poles, and most motorists had to stop to locate the red, white, and blue stripes of the Lincoln, the white-red-white code of the Dixie, or the black-buff-black of the Bankhead. The Lincoln Highway Association, concerned with the problem, claimed that one long stretch of western road carried at one place or another the brands of no fewer than fifteen named roads.

The minor trails were the most fascinating. There was the B V D Trail in Iowa, which had nothing to do with underwear but stood for the Blue Valley Drive and connected Waterloo and Ottumwa; and the Memorial to Daughters and Sons of Republic of Texas Highway, which was only a little longer than its name and ran from San Augustine to San Antonio, Texas. The Crain Highway ran from Upper Marlboro to Dorrs Corner, Maryland, a total of nineteen miles.

Many presidents besides Lincoln got into the act: there were several roads named for Washington, a handful for Teddy Roosevelt, a couple for Thomas Jefferson, and at least one for Andrew Johnson, James Buchanan, Calvin Coolidge, Warren Harding, Woodrow Wilson, Benjamin Harrison, James Polk, James Madison, James Monroe, Herbert Hoover, and William Taft. The Herbert Hoover Highway ran from the Lincoln Highway at Lowden, Iowa, through the Hoover birthplace in West Branch to Iowa City, some forty miles.

A great many highways were named for other notables: war heroes, senators—even literary figures like Rip Van Winkle, whose trail ran from Prattsville to Catskill, New York. Roads like the Cleveland-Marietta-Ashville-Florida Highway told you where they went, and those like the Plank Road in Louisiana, the Hug-the-Coast Highway in Texas, and the Live Stock Route in the South gave travelers some idea of what to expect along the way.

Clearly something needed to be done to aid the traveler and, increasingly, to aid the state road departments which found it harder and harder to grasp a simple identity for a particular road. Some states had initiated infant numbering systems, the simplicity of which suggested that such a scheme ought to be applied to the highways of the country. By 1924, state highway officials from across the country were meeting to discuss a national number system for highways and a uniform style of regulatory and warning signs. The Lincoln Highway Association watched closely as the meetings progressed. It was in favor of a simple numbering system, but like all highway organizations promoting a particular named route, it feared its road would disappear under the shorthand of some arbitrary number, or worse yet, be broken into sections that had one number in one part of the country, another elsewhere. The association wanted numbers that would overlay and stand with the names of the major highways.

L. B. Miller (behind the wheel), C. I. Hansen, and the Wills Sainte Claire pose for a photograph after arriving in Oakland on July 18, 1925, after driving from New York City in four days, six hours, and forty-five minutes. Miller looks ready to turn around and do it again; Hansen looks ready for a long rest.
Courtesy of the University of Michigan.

In March of 1925, the American Association of State Highway Officials suggested a framework for a numbering plan to the secretary of agriculture. W. M. Jardine, the new secretary under Calvin Coolidge, then appointed a board of state and federal highway officials "to undertake immediately the selection and designation of a comprehensive system of through interstate routes, and to devise a comprehensive and uniform scheme for designating such routes in such a manner as to give them a conspicuous place among the highways of the country as roads of interstate and national significance."

As meetings went on in Washington and across the country to decide what routes would be included and what roads would get what numbers, the Lincoln Highway Association sought a hearing with the road officials to insure that the names of the major highways would be retained when the numbers went into effect, and that the Lincoln would be given a single number for its entire length. The officials, however, had decided early in the process to ignore all named roads in their deliberations. If the wishes of one were acknowledged, then all would clamor for inclusion intact. And as discussions among road officials moved into summer, the Lincoln Highway Association, like all road groups, found it had no influence in a process that had states negotiating, scheming, bargaining with every other state that touched their borders, as each tried to acquire and hold onto as many federal routes and miles as possible.

Meanwhile, in the middle of June 1925, the Wendover Cutoff opened with considerable fanfare. At a ceremony out on the salt flats, Secretary Jardine was the featured speaker and complimented Utah on her completion of the difficult section. It had taken nearly eighteen months to complete the forty-mile embankment. Construction had been difficult: it proved necessary to dig a wide trench in the salt flats and fill it with material trucked in from elsewhere in order to provide a stable base for the embankment. Plank seawalls were built along the sides to protect the embankment from wave action. The project had cost $380,000 and made an arrow-straight speedway across the desert.

But beyond Wendover, across northern Nevada, the Victory was still in poor condition. A month after the Wendover Cutoff opened, L. B. Miller crossed on the Lincoln Highway to the south; even though he could have crossed the desert at speed on the new Wendover road, the Victory across northern Nevada was too poor to make good time.

By the fall of 1925, a plan for numbered highways had been hammered out. What finally emerged after considerable wrangling among states and regions of the country was a system that, with the addition of the interstate highways, is essentially unchanged today. On November 19, W. M. Jardine approved the plan calling for some seventy-five thousand miles of numbered federal highways. East-west roads were to be even-numbered; north-south would be odd. Lowest-numbered roads would be in the north and east; the highest, in the south and west. Transcontinental or near-transcontinental highways would bear numbers in multiples of ten—U.S. 10 across the northern tier of states, U.S. 20, 30, and so on to U.S. 90 across the South. Key north-to-south routes would have numbers that ended with 1: U.S. 1 would follow the East Coast from Maine to Florida; U.S. 101 would follow the Pacific Coast from Canada to Mexico. All the federal highways would be identified by the now-familiar federal shield. The individual states would number their own state routes as they saw fit.

The new federal highway system was a near-fatal blow to the Lincoln Highway and a death knell for all the other named highways of the country. Not only was the Lincoln broken up into several numbered roads, but the officials had ruled that all markers and signs for named highways would have to come down. Along with the numbering system, the American Association of State Highway Officials had adopted a much-needed standardization for highway warning and regulation signs, and as part of that package, decided that the signs for the many organizations must come down to avoid hazardous confusion.

Henry Joy wrote Field Secretary Hoag and expressed his bitterness over the decision:

My thought is to send to the president, his cabinet and all members of Congress a copy of the Lincoln Highway Proclamation and along with it a printed slip saying:

"The Lincoln Highway, a memorial to the martyred Lincoln, now known by the grace of God and the authority of the Government of the United States as Federal Route 1, Federal Route 30, Federal Route 30N, Federal Route 30S, Federal Route 530, Federal Route 40 and Federal Route 50."

From Newark to Philadelphia, the Lincoln Highway became U.S. 1. From there west, for two-thirds of its length, clear to Salt Lake City, it became U.S. 30, sometimes splitting for a distance into north and south routes. From there to Ely, it stayed an unnumbered, unofficial desert trail of no significance. It became U.S. 50 across Nevada, assuring federal improvements for the remote highway through Eureka and Austin, then U.S. 40 west of Reno to Sacramento, and U.S. 50 again to Oakland.

U.S. 30 began at Atlantic City, New Jersey, and ended 3,200 miles west at Astoria, Oregon, at the mouth of the Columbia River.

It has been suggested that state and federal highway officials intentionally broke the Lincoln Highway into several numbered routes. The deliberate jog of U.S. 30 far to the north at Salt Lake might suggest that they did. Astoria is 650 miles north of San Francisco, and that terminus gives U.S. 30 a considerable deviation from any straight line. Perhaps the jog to the north was the result of a concern for numbering the Lincoln intact when all the other named roads were broken apart, or lingering bitterness over the struggle for the route in Utah. Or perhaps 30 turned north simply because of the exigencies of state-to-state negotiations and the geographic limitations of road building in the West. Said Joy:

The government, so far as has been within its power, has obliterated the Lincoln Highway from the memory of man, in spite of the fact that the press of the United States gave to the work very many hundreds of thousands of dollars worth of publicity to aid in putting this wonderful main arterial highway to the memory of Abraham Lincoln into actual service for the people of America.

By 1926, attendance at Lincoln Highway Association board and annual meetings had declined. With the failure in Utah and the new federal numbering system, interest in the Lincoln Way had dropped off on the part of the officers, directors, and the public. Operating finances for the association were growing tight. It would now be up to the government to build roads for the country. The Lincoln Highway Association, by pushing for its road as the first of a national system of roads, had promoted itself out of a job.

On the evening of November 11, 1927, the board of directors of the Lincoln Highway Association met at the Detroit Athletic Club. Among those present were men who had been with the group from the start: Frank Seiberling, Henry Joy, Austin Bement, Roy Chapin, and Gael Hoag. President J. Newton Gunn had been ill much of 1927 and was not present. Neither was Carl Fisher, the dreamer, promoter, and visionary who had set the imaginations of these men to work for highways in America. The minutes of two earlier meetings were approved, then a motion was made that the Lincoln Highway Association cease its "active and aggressive operations" as of December 31, 1927.

Though annual board meetings would continue for another few years, the offices in Detroit would close; the motoring public would no longer be able to buy guidebooks, decals, and pennants, or get school material relating to Lincoln, good roads, or the geography of the nation along the Lincoln Highway. Consuls and supporters across the country urged the association to continue, but the work was done; America was getting her transcontinental highways. The inactive association would now merely do what it could to see the name Lincoln Highway perpetuated.

J. Newton Gunn died on November 26, 1927, and Frank Seiberling found himself president of the Lincoln Highway Association for the third time. Though he took the presidency under protest, he continued to believe strongly in the highway and what it stood for. As the offices of the association were being cleaned out, records moved, and furniture sold, Seiberling and the board worked on two pieces of unfinished business: some resolution to the missing highway in Utah and a permanent marking of the route as a memorial to Lincoln.

In exchange for an agreement to build a north-south connecting road between Wendover and Ely, a 122-mile link that would connect the Victory with the Lincoln, the association very reluctantly agreed to abandon its route across Utah and to decree the rival Wendover desert crossing and the new link as the official Lincoln Highway. Henry Joy must at last have seen his error in the original selection of the route in Utah and Nevada; he suggested that the association move the Lincoln to the now-main route through Elko and Winnemucca as well, but the board refused. In its motion to approve the Wendover road and the link between there and Ely, the board expressed the hope that the much-fought-over Utah road would be improved eventually, but everyone, especially Henry Joy, knew it never would be.

At about the same time, a new state was added to the route of the Lincoln Highway—West Virginia. The route between Pittsburgh and the Ohio line had followed a slow, narrow route through numerous small industrial towns along the north side of the Ohio River. It often took three hours of stop-and-go driving to cover the fifty-two miles west of Pittsburgh. The association had pushed and finally won approval for a federal route directly northwest and overland from Pittsburgh to Newell, West Virginia, where the highway could cross the Ohio to East Liverpool, Ohio. This route shortened the highway by

thirteen miles and cut the driving time in half. West Virginia now proudly hosted the Lincoln Highway for something less than five miles.

By the end of 1927, the old trail markers were coming down as fast as the new federal shields were going up. The painted bands on telephone poles were fading, and travelers were beginning to forget the names of the old roads as the easier-to-use numbers came into use. Although the American Association of State Highway Officials had banned the maintenance and erection of signs or markers for named roads, the Lincoln Highway Association petitioned and was granted permission to mark the highway. It won the approval based on the idea that it wasn't planning to mark the highway as a road, as a route from one place to another, but as a memorial to Abraham Lincoln. And rather than banding telephone poles or painting gaudy signs that might distract the motorist, the association proposed small concrete posts to be set some distance from the roadway.

In the summer of 1928, Gael Hoag, the last paid representative of the Lincoln Highway Association, took the organization's open-top Packard touring car and made the last official coast-to-coast trip on the Lincoln Highway. He was one of three men who knew or had known every curve, hill, vista, and mudhole along this highway. Like Henry Joy and Harry Ostermann before him, Hoag knew this path from sea to sea as well as most people know and cherish a treasured poem or Bible verse. Hoag, Joy, and Ostermann were unusual even within their own organization: Hoag once estimated that probably no more than one-fourth of the officers and directors had ever traveled the whole highway.

Hoag made detailed notes about where the markers were to be located—including along the new Wendover route—visited for a last time with many consuls, shook hands, and said good-bye.

With the help of old friends in the cement industry, some three thousand markers were cast, each with a small directional arrow and a small bronze bust of Lincoln. Around his profile were the words "This highway dedicated to Abraham Lincoln." It was a last flourish for the Lincoln Highway Association, the last of many efforts that was to keep this highway in the minds of Americans across the nation, and it was planned accordingly. Gael Hoag arranged for the Boy Scouts to place these markers along the highway, and better yet, arranged with troops across the country to have them all placed on the same day.

The markers were shipped to towns and cities along the line, and the holes were dug in preparation. Then, on September 1, 1928, eight months after the active association had ceased to exist, Boy Scout troops across the country fanned out with their loads of markers. At an average of nearly one per mile, they lowered the concrete markers into the holes, leveled them, tamped the soil tight around them, and went home.

At the close of the 1920s, the average new car had four-wheel brakes and smooth-riding, puncture-resistant balloon tires, and stood much lower than cars ever had before. Packard had developed a rear differential that used quieter hypoid gears and allowed the drive shaft to be lowered, and the chassis and body as well. In addition, cars no longer needed to be built low to operate on rutted, unimproved roads. Most automobiles now spent their lives on 800,000 miles of improved highways. Cars were quieter, easier to handle, safer, and more reliable. Heaters and radios were available in several models. New York hotel lobbies bulged with display cars during the National Automobile Show of 1929, as throngs of shoppers looked, compared, and bought. Production of motor vehicles broke all records that year: better than five million new cars, trucks, and buses took to the highways. Not for twenty years would that record fall.

In 1919, ten years earlier, nine out of ten new cars had been open touring models. By 1929, the ratio was reversed: for every new open car sold, nine hard-topped sedans or coupes went to eager owners. The new breed of motorist wanted comfort; they were tired of the rain, the cold, the insects, the dust.

And they were tired of the wind.

9

Going My Way on the Lincoln Highway

Hi—

Stayed here all nite, and writing this on my lap as we drive along, the road has been wonderful. Three years ago we stayed at Fremont, Nebr. the first night so made a few more miles this time.

Bye for now,
Ralph and Effie
—Postcard from
Keen Korner Cabins,
Hi-way 30, Columbus,
Nebraska, ca. 1940

While people soon forgot about the Victory, soon forgot about the Midland Trail, the Golden Rod Hi-Way, the Waubonsie Trail, and the Red Ball Route, they didn't forget the Lincoln. This highway, though no longer given any official significance, remained in the public mind as a symbol of freedom, patriotism, and the American wandering spirit.

And it wasn't the thin line of small concrete markers that kept the Lincoln Highway alive; its longevity was a result of the grand publicity campaign that Henry Joy and the association managed so well; an entire generation of Americans knew the Lincoln Highway and the states it touched. "New York—New Jersey—Pennsylvania—Ohio—Indiana—Illinois—Iowa—Nebraska—Wyoming—Utah—Nevada—and—California!" was still a chant for kids in grade school geography.

Henry Joy once said that he felt every dollar spent by the Lincoln Highway Association was spent for education—that money spent on seedling miles, guidebooks, newspaper space really bought an educated public that wanted good roads. Certainly the association had been no road-building organization; the only construction tool it ever owned was a grader it purchased and loaned to a Utah rancher so he could keep the Goodyear Cutoff open.

During its fourteen-year active life, some $1,250,000 passed across the books of the association. A bit more than half was used for salaries, rent, printing, postage, advertisements, and promotion. The smaller portion went on to finance road building. To this private cash was added donations of concrete and other materials, and local, state, and federal money that brought measurable road improvement to the Lincoln Way. Much of this work was swallowed and graded over in the rush to pave America later, but the efforts of the association and the state and county crews that did the work eased travel difficulties until the federal government began to fund road building in a major way.

The United States would have eventually gotten paved highways even without the example set by the Lincoln Highway. The democra-

1915 Lincoln Highway through Burns, Wyoming.

At greater intervals you come to towns and you drive between two closely fitted rows of oddly assorted domino-shaped stores and houses, and then on out upon the great flat table again. For scores and scores of miles the scene is unvarying. On and on you go over that endless road until at last far, far on the gray horizon you catch the first faint glint of the white-peaked Rocky Mountains.
—Emily Post, 1915

Bosler, Wyoming.

Red Top Service, Rock River, Wyoming, on the Laramie Plains.

Sign for Craig Cafe, Rawlins, in Carbon County, Wyoming.

Yet even on these lonely stretches modern civilization had supplied constant reminders that life is short and chancy. At intervals here and later, hundreds of miles from no-where, we glimpsed the wrecks of auto-mobiles, bleached bones of little cars that died to make men free.
—Mary Day Winn, 1931
  The Macadam Trail

U.S. 30 / Lincoln Highway and Interstate 80 near the Continental Divide, Creston, Wyoming.

On the top of the Great Continental Divide we stopped to eat our lunch. We had approached it so gradually that we did not realize we were there until we saw the monument marking the grave of Frank Yort, the surveyor who determined the line. We thought it a rather lonely spot either to start a cemetery or to await the Judgment Day.
—Bellamy Partridge, 1913

Monument to Henry Joy, Creston, Wyoming.

All day long we traveled through prairie land, framed in distant mountains and filled with magical alternations of sunlight and shade. There is a spell in the grand emptiness of Wyoming. Imperceptibly, it creeps into you and clutches you. Once you have journeyed through its wide-flung brown wastes, forever after you long to return to them.
—Frederic Van de Water, 1927

Green River, Wyoming.

Green River Town is located at the junction of the road and the river. During the summer months travelers forded the river, but we were curious as to how the river could be crossed in the winter-time. When we asked this question, we received a very simple and logical answer—"No one travels during the winter."
—Thornton Round, 1914
  The Good of It All

Union Pacific Railroad and Lincoln Highway, Echo Canyon, Summit County, Utah.

The sun was getting low but there was still enough light to see the sights and even to take photographs as we drove through these massive wooded mountains. We liked the change from the barren scenery of Wyoming. One side of the pass would be bathed in brilliant sunlight while heavy shadows engulfed the opposite side, making the scene appear almost black-and-white in dazzling contrast.
—Alice Ramsey, 1909
  Veil, Duster and Tire Iron

Weber Valley, Wanship, Utah.

Simpson Buttes, Tooele County, Utah.

Site of J. J. Thomas Ranch,
Fish Springs, Utah.

We came to Fish Springs Ranch in the midst
of this lonely country and stopped for
luncheon. Our host was a tall and power-
fully built elderly ranchman in a blue
jumper.

"Horace Greeley ate at this table when he
came on his historic Western trip, and so I
keep the place standing," he said. His young
helper cooked our meal in the back room
and our host served it in the front one. We
had fried eggs, potatoes, pickles, cheese,
bread, butter, and tea, and an appetizing
cup cake cut into square pieces. Our host was
very hospitable. "Have some of them sweet
pickles, folks."
—Effie Gladding, 1914

Pony express route, stage coach route, and
1915 Lincoln Highway near Fish Springs,
Utah.

The tan stretches of sand, the dusky patches
of sage, the false frost of alkali through
which the rejuvenated trails lead have not
changed. Long miles of prairie still lie as
bare, as untamed as when the first covered
wagons lurched across them. Rubber tires
now stir the dust through which the ox
teams plodded. The miles that tourists drop
behind them in five hours would have taken
a week to traverse in the old days. Nightfall
brings, to-day, not a defensive circle of wag-
ons, but tents that rise in an auto camp.
—Frederic Van de Water, 1927

Great Salt Lake Desert,
Tooele County, Utah.

As we slowly dropped down the mountain to the Desert we passed a red sign board that warned those about to make this trip to be sure they were plentifully supplied with gas, oil, and water!

After winding down for a half hour we struck the straight trail that led for twenty miles across the alkali desert. Thunder storms with grey green clouds filled the sky ahead, making a picture of Purgatory for even the most exacting. A few drops fell on us once in a while when the sun was scorching overhead. All I can say is I was vastly pleased to get across that nightmare. The mere recollection of it makes me so thirsty that I have to stop and get several glasses of ice water!
—James M. Flagg, 1925
    Boulevards All the Way—Maybe

Near the Goodyear Cutoff, Tooele County,
Utah.

Gold Hill, Utah, on the 1926 Lincoln Highway.

tization of the road came with cheaper cars and the working person's desire to own one. The country simply could not go long with millions of cars and dirt roads; the change was inevitable. What the Lincoln Highway Association did was speed up the process; it made the public accept the idea of long roads built not for local convenience but for the benefit of everyone—roads for the nation, not just the county. This was the Fisher idea, a road for the whole nation as an example to all and as a path for all—a democratic ideal that lived on in the words of songs, the names of little hotels and streets, a chain of concrete markers, and the restless psyche of America.

During the summer of 1929 Gael Hoag left Detroit for his new home in Oakland after the final business of the association was tied up. The association had retained Hoag for a number of months after the offices closed. There had been final business to attend to and some difficulty with the markers in Wyoming. Hoag drove the Lincoln Highway Association Packard, but it was his car now. Seiberling and the executive committee had voted that the car be given to him. The notes of the meeting state, "In moving this gift, Mr. Seiberling stated he felt Mr. Hoag had handled the signing matter and the closing of the Association's affairs in an excellent manner and that the gift of the car, valued at $500.00, might be considered a token of appreciation." The car would become Hoag's at the Utah state line, after he had seen to the markers in Wyoming.

This Packard Eight was a proud anachronism. It had come to the association in 1924, brand new from Packard, a heavy and solid open touring car with narrow windshield, giant headlamps, classic Packard radiator front, and a long louvered hood to cover the in-line eight-cylinder engine. Gael Hoag loved this car. He had proudly claimed that some of its special equipment—the latest and largest six-ply Fisk balloon touring tires, Anderson Spring "Spats," and Lovejoy shock absorbers—made it "the fastest and most comfortable touring car on the road." It carried Lincoln Highway emblems on the doors and bands of red, white, and blue clear around the machine, "to help consuls identify the car if they should meet it on the road."

The odometer registered the thousands of miles that Hoag had driven behind the wheel of this car. He and the car had crossed the continent together many times. This time, in addition to his luggage, the back of the big Packard Eight touring car probably carried boxes of business records, correspondence, field notes, and unsold guides on to Oakland, where Hoag would continue to act as secretary of the inactive Lincoln Highway Association.

By 1931, federal money had turned all but a few miles of the Lincoln Highway into an improved, all-weather highway. Even in the old mud state of Iowa, travelers no longer made lengthy unscheduled stops during the wet months. The political climate had changed as farmers bought more and more cars and the last of the unfair adjacent-owner taxation laws had been repealed, and Iowa was pouring paving to catch up. Unbroken paving extended from New York to Missouri Valley, Iowa, and about half the remaining distance to San Francisco was also hard-surfaced. Only about thirty miles in Nebraska and Wyoming were unimproved, having no embankment, drainage, or paving.

L. B. Miller made his last-known record run that year, another coast-to-coast-to-coast round trip, this time in a Plymouth. He left San Francisco on August 4, reached New York on the seventh, where he spent just over an hour, and arrived back in San Francisco on August 9. His elapsed time was five days, twelve hours, and nine minutes. Soon after his return to California, he faded from sight and memory. Miller the speedster had left a few obscure marks in the record book, but his decade was finished. By 1931, the days of automobile record making and record breaking were over. America had turned to her new speed passion, the airplane. The sky was full of unmade records, but the road was filled with automobiles.

Henry Bourne Joy died on November 6, 1936, at his home in Michigan. He was seventy-one. His association with Packard had ended eighteen years earlier, but his Lincoln Highway interests continued to the end. Though it is doubtful that he drove the highway anytime after 1920, he had still held a strong love for that path he knew so well and had pushed so hard to complete.

On the dome of the continent where miles of sagebrush and sky greet the traveler, right where the Lincoln Highway crosses the continental divide 194 miles west of Cheyenne, Henry Joy's family erected and dedicated a monument to him. Perhaps it was near that place where Joy, Bement, and Eisenhut had camped under the Wyoming stars after two weeks of rain and struggle crossing the Midwest in 1915. Though not a very scenic place according to usual tastes, it is a landmark of the continent in a part of the country that Henry Joy loved above all others. Travelers speeding past this place on U.S. 30 on July 4, 1939, must have wondered at the little ceremony beside the road. So far from anywhere, a small gathering of family and friends made dedications and thought of the man who had passed here so many times. Joy had once said, "I consider the Lincoln Highway the greatest thing I ever did in my life."

At four corners to an inscribed stone tablet are concrete Lincoln Highway markers, guarding the monument and standing as corner posts for an iron fence. At greater distance are two other 1928 markers, outlining the perimeter of a turnaround and standing as advance guards against the Wyoming wind. The tablet quotes Joy: "That there should be a Lincoln Highway across this country is the important thing."

And what of Carl Fisher and Miami Beach? The built-up sandspit took a little longer to catch on than did the Lincoln Highway, but if Carl Fisher was good at anything, it was that he could spot a good idea, a future trend, and be right there on the ground floor when it took off. Fisher had staked his substantial fortune on this mangrove-swamp-turned-sandspit, and by 1922 the whole project stood on the brink of ruin. The skeptics were ready to say I told you so when, all at once, land began to sell, the Dixie Highway began to bring snowbirds, and Fisher began to get very rich. He built golf courses, polo fields, yacht moorings, shops, and office buildings. Fisher built four hotels in Miami Beach at the end of the Dixie Highway: the King Cole, the Flamingo, the Nautilus, and the Lincoln. Will Rogers once referred to Fisher as "the midwife of Florida. Had there been no Fisher, Florida would be known today as just 'The Turpentine State.' He rehearsed the mosquitoes till they wouldn't bite you until after you had bought." Some estimate Fisher was worth $100 million by 1925.

Success breeds success, and Fisher turned to another project: a similar but even larger vacation city at Montauk Point, on the tip of Long Island. It would be a summer playground and a harbor for transatlantic passenger liners. The steamers, Fisher calculated, could save a day by docking at Montauk instead of going clear to New York City, which, by rail, was only four hours away.

He bought about ten thousand acres and set to work on an even larger scale than in Florida. With his Miami Beach holdings to guarantee construction bonds, Carl Fisher began building roads, hotels, and a deep harbor. He was directing the work of an army of men when the great hurricane of 1926 struck Florida. "MIAMI BEACH TOTAL LOSS," his agent wired. The house of Fisher began to tumble, and with the market crash of 1929 and the ensuing depression, he was wiped out.

He spent his last decade in Miami Beach, in the rebuilt city that was no longer his, not quite in poverty, but with none of the monied flamboyance that had characterized the earlier Fisher. He was often seen walking the beach in his Norfolk jacket, white flannel trousers, and floppy felt hat, talking of great plans for the Florida Keys. Carl Fisher died in July 1939.

Gael Hoag and the Lincoln Highway Association Packard, 1926. A symbol of the association, the powerful car represented *grace, pride, speed, and things done the right way. When it became Gael Hoag's in 1929, this car—like the association itself—* *represented something whose time had passed. It was an open touring automobile from a grand tradition, but by then most* *new cars were enclosed and all the roads had numbers instead of names.*
Courtesy of the University of Michigan.

Sometime between 1915, when Emily Post stuck fast in the mud of Illinois, and the early 1930s, when the Lincoln Highway finally became an improved highway from coast to coast, the adventure of auto travel began to fade. Certainly the novelty had worn off; it was no longer noteworthy or newsworthy to make a long auto trip. A few people still wrote books about transcontinental journeys, but because travel was now so easy, the authors concentrated on social concerns or personal growth, or used the highway as a metaphor for American life and politics. The road and the landscape had become window dressing to another story.

By now everybody had a car or was planning on getting one soon, and thousands drove autos and trucks for business and errands, accumulating dull mile after dull mile. Especially in the East, heavy traffic also tempered the thrill. No one could feel much like a pioneer, a pathfinder, when trapped in a stationary car in rush hour traffic.

To a certain extent, the new highways themselves were to blame. Good highways brought more autos to the road, clogging the way. The hard-won paved wonders that all had boasted would free people to explore America had in fact freed them from exploring the country. Increasingly they wanted to get from place to place as fast as possible. The better the road, the faster drivers traveled; the faster they went, the less they saw along the way, and the less they cared to see. Getting through and avoiding traffic became the greatest goals. Main Street, stops for gas and oil, winding roads, and intersections were all becoming inconveniences somehow as the trip became less the focus than the destination. *Fortune* magazine summed it up well in 1934: " . . . the habit is upon us to refuel and eat and sleep and amuse ourselves not in the towns as towns—they slow us up—but along the open roadside which is a new kind of town in itself, and in the little towns that have all but turned themselves into roadsides."

Myron Stearns drove across the country in 1926 and again ten years later in 1936. On the earlier trip, with stops for gas and infor-

mation, and slowdowns for detours, cities, and what he called "similar inconveniences," Stearns averaged twenty miles an hour. In 1936 he averaged forty. On this trip he boasted that the car he drove west could cruise at nearly seventy miles per hour uphill and down, and by that time, a few stretches of a few roads were good enough to try it. "Four hundred miles a day, over open highways, was easier than two hundred had been ten years before. Some days we made five hundred. A week from coast to coast."

Though L. B. Miller had faded from the scene, motorists adopted his style, boasting of the quickest run from one place to another. Speed became the measure of a trip, not adventure or enjoyment along the way. Said motorist Stearns: "Floods, tornadoes, depressions, Old Deals and New Deals, may come and go—but year after year the stream of traffic, bigger and busier than ever, rolls two miles an hour faster. . . . Through speed-limits and safety campaigns, grade-crossings, traffic laws, stop signs and motor cops—two miles an hour faster." The tense grasp of the wheel with hunched shoulders became the posture of the auto traveler.

Certainly no one missed being stuck in the sand in Wyoming or pined for the days of breakdown far from town, but with great road improvement and enclosed cars with radios came isolation from the land around. Seen through the safety glass of a Plymouth, the world became as remote as it had been through the plate glass of a Pullman.

What Joy, Fisher, and thousands of other automobile- and road-minded individuals began in the first decades of the century blossomed after they were gone. On the now-paved highway were cars that had reached a high level of utility, and beside the highway, roadside America was rising. Americans had found a new mode of travel and were now busy creating a landscape to support it.

The 1930s and 1940s were the embryonic years for our now-familiar roadside. While the twenties brought enormous numbers of vehicles, it took a little longer for the roadscape to catch up. In 1925, an oldster could still take the grandkids back to the hometown and show them childhood landmarks—the roadside orchard, a stream ford used by the covered wagons, the old commercial hotel and dining room, the tallest building in town. The traveler of those days saw a landscape that was little changed from the way it had looked fifty or even a hundred years earlier. A filling station had replaced the livery stable; a roadside eatery and a tourist camp had sprung up, and a night spot or two had appeared. But along the narrow and now paved highway, the landscape showed only the faint hint of what was to come.

It was still a simpler looking America, a place that felt like what home was supposed to be. It was a place where the road was under the power of the landscape, where the pavement followed the form of the earth, and where the nonhighway landscape—that of farming, manufacturing, the railroad, main street, and neighborhoods—still began right at the edge of the concrete.

But the signals of change were there. When the cafe at the edge of town paved a large parking lot to accommodate trucks, when the farmer began to sit on the back porch instead of the front because of the road noise, when the highway department added a two-foot section of concrete to each side of the highway for the wide trucks, when the owner of the tourist camp built a row of little huts, complete with trellises, the great change in the American landscape had begun.

The American gas station may have been the leading indicator of the stampede to the roadside. In the opening years of the century, gasoline had been sold from cans or from curbside pumps in front of the general store or livery stable. By the 1920s, the filling station— a new sort of business dedicated entirely to the auto—had become a common sight. Filling stations most often were small, house-shaped buildings barely large enough for an office and heating stove, with a canopy and high, glass-topped pumps. Where possible they were built on street corners for easiest access by automobiles. From the beginning, neighbors complained about the noise, the lights at night, and the disruption.

For many years, the businesses of refueling cars and repairing them were separate: filling stations took care of gas and oil, garages did the mechanical work. But by the early 1930s, many new stations had appeared that combined the retailing of fuel and oil with indoor bays for repair. The quintessential box service station appeared, a pattern that would be the standard for the next four decades. As these stations grew thick in villages and cities and spread out along the highway to the edge of town, they became larger and paved greater chunks of land. They crowded into neighborhoods of all sorts and butted up against storefronts and shaded Victorian homes.

During the depression, oil companies had to work harder for sales in a temporarily shrinking market, so stations strutted the latest in colors, designs, and services, including the breathtaking streamlined look. Some station owners tried outlandish architecture to draw the customer off the road and to their pumps. Stations built to resemble English country homes, oil cans, Indian tepees, Dutch windmills, lighthouses, and any other imagined symbol or object appeared along the road in the 1920s and 1930s.

Side by side with the gas stations were new eating establishments that had sprung up along the highway. No longer did the hungry motorist need to seek the hotel dining room downtown: lunch stands, cafes, diners, roadhouses, tearooms, family restaurants, and drive-ins all catering to the motor trade appeared along the highway everywhere. These new places turned away from the railroad and town center; they didn't care about the train passenger or local businessman or businesswoman; they were dedicated to luring the passing automobile traveler.

The variations were many. Some began as lunch or barbecue stands, others as diners, which had originally been downtown fixtures but by the 1930s were being moved whole and intact out along the highway.

A few diners were actually adapted railroad or trolley cars, but most were constructed in factories as eating places from the start. They were moved into place like house trailers, and truckers and salesmen could be hunched over ham and eggs only a day or two after a diner had arrived at the site. They were sleek, impressionistic visions of travel, specifically mimicking the railroad; but with their bright stainless steel and neon, they quickly became emblematic of the highway.

After prohibition ended, roadhouses—some were old tearooms, some were new buildings, some even were old houses—appeared along the highway at the edge of town or in the countryside, flashing beer and cocktail neon into the night. Often they were set back from the road and surrounded by trees, adding a mysterious, secretive touch. They were supper clubs, bars, and dance halls, catering to locals and travelers alike. Local bands strove for the big-band sound and filled the night with music. They added color to the roadside and often added drunk drivers to the highway, a new hazard of auto travel.

As if to undermine the slightly seedy impression of roadhouses and highway bars, family restaurants began to elbow in along the ever more crowded roadside. These were places that offered much broader menus than the average diner or cafe, boasted better food than a greasy spoon, and sought to attract the traveling family, not just the truck driver or businessman. The leader in the field was Howard D. Johnson, who, by 1940, had parlayed his oceanside ice-cream stands into 125 family restaurants along the East Coast from Maine to Florida.

The early drive-ins and franchise fast-food emporia soon took up their appointed place by the road. This new sort of eating geared to speed and convenience had its roots in a couple of earlier traditions. First was the food stand that sprouted along the road after World War I. These hot dog stands often were informal shanty affairs that sometimes served remarkably inferior food and earned the moniker "Ptomaine Joe's."

The other phenomenon was the standardized-fare hamburger shop—namely White Castle. In order to overcome the poor reputation of the usual food stand, White Castle came to the market selling a standardized high-quality product that the customer could watch being made. The menu had few options, but with a predictable level of quality better than the average food stand, brought customers from the start. Others began to copy the White Castle equation with even greater success. The first White Castle hamburger stand opened in 1921, and A&W introduced "tray girls" in 1924; together these businesses planted the seeds for food-franchised America.

But before the franchises took over, before the interstate, Americans on the road ate at another, more common institution. In no way standardized, it was plain and ubiquitous. It wasn't homey enough to be a tearoom, not shiny enough to be a diner, not big enough to be a family restaurant. It usually announced itself unmistakably and unpretentiously with a homemade sign of four big letters: CAFE.

In the earlier days, before the highway moved to a bypass, many cafes were downtown and Main Street places that simply and gladly accepted the increased business that auto travelers provided. Others were places that had moved away from downtown into residential areas or out to the edge of town to cater expressly to the motoring hungry.

Each was unique, each was comfortably familiar. There were often Coca-Cola bottle-cap signs nailed to the outside, curtains in the windows, chairs and tables that seldom matched. Decor and menu were a matter of the tastes of the owner—sometimes with some sort of ethnic flavor, other times quite basic. Some places were shiny and efficient, like diners, while others might reflect a local theme. A few bore the artistry of itinerant artists who, in order to pay for their meals, painted murals of fantastic wilderness scenes—mountains, forests, and lakes that were landscapes of imagination rather than reality.

Menus all bore a family resemblance: hamburgers, cold sandwiches, eggs, and the blue plate special at noon.

Though in no way standardized, cafes usually had certain items in common: on each table stood on ashtray with the name of the local garage printed on the bottom, a bottle of ketchup, a napkin holder, salt and pepper, and a creamer. There was a calendar or two on the wall from the local cattle buyer or the insurance agent, and a big urn making gallons of coffee, hustled by waitresses to customers: farmers, local business people, a delivery man on his lunch break, families, and travelers. Meals came on scratched old plates, with the trim glaze worn away from thousands of washings. The cups, too, were worn, and usually there was a puddle of coffee in the saucer because the waitress was in a hurry. In the days before fast food, this is where most Americans ate when they didn't eat at home.

It wasn't long before someone crossed a row of gas pumps with an eating place and bore the first truck stop. Like roadhouses, truck stops were often out in the hinterlands, or at least at the edge of town, where land was cheaper. To capitalize on every possible customer, they were usually built at the intersections of major highways. The basic criterion was a parking lot big enough for ever-bigger trucks. All that pavement might surround something resembling a diner, cafe, or family restaurant, but when the rigs nuzzled up to the windows like cattle in a storm, the place became a truck stop. As the first urban bypasses drew arcs around cities, truck stops sprouted like cattails in roadside ditches.

As the great American roadside grew into an institution, entrepreneurs invented new places for travelers to rest from a long day on the road. Tourist camps became cabin camps, motor courts, then motels. They had begun as simple campgrounds in the days following World War I: shaded grassy areas, usually owned by the city or town, and usually free, where travelers in rough clothes pitched tents, cooked

With a name more likely for a revolutionary war–era tavern, the Cross Keys Diner along the Lincoln Highway in New Oxford, Pennsylvania, catered to travelers of all sorts. Though it looks to have been built of several railroad cars, it probably was a prefabricated structure set up on site.
Courtesy of Lyell Henry.

AIR CONDITIONED CAFE

KING TOWER
½ Mile East
TAMA, IOWA

MODERN HEATED CABINS

TAMA, THE HOME OF
IOWA'S ONLY INDIAN
RESERVATION

Open 24 Hours

GAS

CABINS

KING TOWER INC.

HIGHWAYS U.S. 30 & 64

LOCATED HALF-WAY BETWEEN CHICAGO AND
OMAHA OR SIOUX CITY, IOWA

The "King Tower One-Stop" of Tama, Iowa,
boasted twenty-four-hour service and was a
combination cafe, gas station, and tourist
court.
Author's collection.

WAY FARERS CAMP ON LINCOLN HIGHWAY-BELLE PLAINE IA

The Way Farers Camp, a mile east of Belle Plaine, Iowa, represents the first evolutionary step from open tourist camps where travelers pitched tents to the cabin camp, where travelers rented cabins just big enough for a bed and washstand. Author's collection.

over fires, and kept lookout for road liars. When communities began charging for overnight stays, private operators got into the act, and soon competition had them all adding amenities like laundry buildings, picnic tables, hot showers, even electric hookups.

But as travelers turned away from riding in open cars, they also turned away from the notion of sleeping in open tents and roughing it in general. Operators soon began building tiny cabins so travelers could leave the tent at home. At first these cabins were little more than chicken coops, hardly big enough for a bed, but travelers loved them because they still provided the feeling of camping, of the outdoors, yet added a modicum of privacy and protection from the elements. Day by day these "cabin camps" added the comforts of home: bedding, electricity, heating, kitchenettes, and before long, camping was forgotten. A writer for *Fortune* described a typical cabin in 1934:

. . . you find a small, clean room, perhaps ten by twelve. Typically, its furniture is a double bed—a sign may have told you it is a Simmons, with Beautyrest mattress—a table, two kitchen chairs, a small mirror, a row of hooks. In one corner a washbasin with cold running water; in another, the half-opened door to a toilet. There is a bit of chintz curtaining over the screened windows, through which a breeze is blowing.

Compared with the difficulty of getting to the downtown hotel through "ten thousand impediments to motion," and the rather formal atmosphere of even second-class hotels, the casual cabin camp made it easy for the traveler to bed down, without having to tip bellhops and doormen.

These establishments grew into motor courts, and the cabins were often arranged in L or U shapes with a central courtyard to add to the sense of security and coziness. Sometime in the late 1940s, the term *motel* came into common use. By now the cabins had been connected, neon added wherever possible, and the motorist slept in a miniature, idealized version of home, complete with oil paintings on the wall, a Bible, and hangers in the closet. The dirty-fingernailed running-board campers of the past were long forgotten.

The flood of new business of all kinds along the American highway brought a great wave of roadside advertisement and attraction. Competition grew stiff as new places came to the highway and as America entered the depths of depression. Businesses of all sorts used gimmicks of any kind to get the traveler to turn off here for ice cream, coffee, barbecue, gas, a night's rest, or a stuffed-animal museum. Buildings themselves were advertisements when the ice-cream stand was built to resemble an open carton, or when the lunch stand was constructed as a big coffee pot, or when giant concrete animals—fish, dinosaurs, bison, alligators, snakes—gave visual testament to the thrills to be found within.

The earliest roadside advertising of the auto age had continued the traditions of the nineteenth century—logos and slogans painted on buildings and simple signs nailed to trees. In most rural areas of the country, barns were common, and inasmuch as a farmer was often happy to get a free paint job and a case of the product in question once in a while, he often let the painter from Bowes Seal Fast or Fisk Tires or Mail Pouch Tobacco paint the side facing the highway. Lucky indeed was the farmer who had a hilltop farm with barn in prominent view from either direction, who could rake in cash for a prime location as well as a free paint job.

It didn't take people long to see other possibilities. Even as the signs for the named roads were being taken down, the signs hawking countless products—both automotive and not—attractions, and roadside businesses were appearing everywhere. In places near cities and resort areas, the roadscape became a tunnel of billboards, masking the landscape from view. Dense along the highway, they announced "DE LUXE CABINS ONE MILE," "VISIT WALL DRUG," "GRAND VIEW POINT HOTEL—SEE 3 STATES AND 7 COUNTIES."

Perhaps the most endearing and enduring roadside ads were the little Burma-Shave signs scattered across the country. They stood by the road in a series of five or six small boards about a hundred paces apart and carried abbreviated verse extolling the virtues of Burma-Shave brushless shaving cream. The meter of the verse was set by the signs, and the last in the series always carried the name of the product: HALF A POUND / FOR / HALF A DOLLAR / SPREAD ON THIN / ABOVE THE COLLAR / BURMA-SHAVE. They were happy surprises for travelers from coast to coast, as whole carloads would read them aloud in unison. BENEATH THIS STONE / LIES ELMER GUSH / TICKLED TO DEATH / BY HIS / SHAVING BRUSH. In any carload of travelers, one person was given the duty of deciphering those passed in the opposite direction: ANOTHER SALE / RINGS UP / SOME STORE / WITHOUT FAIL / EVERY SECOND.

And sell they did. The little signs first appeared in 1925, and within a few short years they had taken a nearly bankrupt company to annual sales greater than three million dollars.

The burgeoning American roadside flourished, grew in breadth and depth despite the hold of the depression. According to author Lloyd Morris, this new place became an "endless concrete emporium," a "transcontinental bazaar" that offered the traveler "lawn furniture, 'artistic' weathervanes, sundials and garden sculpture, assorted antiques, boxes of nuts assembled from the ends of the earth, golf hats and clubs, fishworms and tackle, picture postcards of places he had passed without seeing, domestic pets." Another writer estimated that gas stations, camps, cafes, food stands, resorts, and attractions would gross nearly three billion dollars in the hard year of 1934.

During the depression, sales of new cars hit rock bottom, but gasoline sales dipped only slightly. People would give up the farm, give up the furniture, but never the car. While many who were down and out turned to the highway to take them to milk and honey in California, others turned to the highway to make a buck, opening fruit and vege-

table stands, cafes, wondrous caverns, hot dog stands, wax museums, filling stations, drive-ins, and motor courts.

The Lincoln Highway, like other major roads in the United States, became the seedbed for countless entrepreneurial ventures. Most started out small and some stayed that way, while others flourished and grew, and some died on the vine shortly after the doors were opened. A good many of these places have disappeared without a trace, and others have been made over so many times that it is hard to imagine that what is now a one-hundred-table family restaurant began as an unpretentious hot dog and soft drink stand.

The histories of these places can often only be learned from the extant postcards that they handed out or sold by the thousands during their lifetime. Through these cards we come to know the Kampus Camp in Ames, Iowa; the Keen Korner Cabins in Columbus, Nebraska; and making fuller use of the alliteration, the Kearney Kabin Kamp in Kearney, Nebraska. Some, instead of having cute names, relied on the name of the owner. Wally's Hi-way Cafe in Lexington, Nebraska, boasted the ubiquitous slogan "Home cooked meals" and called itself "A Tourists' Oasis."

"Doc" Seylars Rest House stood on Tuscarora Summit in Pennsylvania, boasted an elevation of 2,240 feet, and advertised lunches, ice cream, souvenirs, "real cigars," and "surpassing coffee." Tourists stopped for all those things plus a chance to climb to the viewing platform on top of the lunchroom to have a look around the mountains of Pennsylvania. Although Doc exaggerated the altitude by about 120 feet, the view was among the best in the mountains.

Eighteen miles west of Doc's was another scenic view at Bill's Place. This stopping place stood on Ray's Hill and no doubt competed with Doc for the same motoring trade. At 1,958 feet, it also had a viewing platform, but not on the lunchroom. Bill's dispensed gasoline from an array of no fewer than ten different roadside pumps, and while the tank was being filled, a tourist could browse through the open-air souvenir stand and buy rough-hewn stools, tanned hides, hanging flowerpots, postcards, and look at the Lincoln Highway marker by the door. Bill bragged of the highest and smallest post office in the U.S., but Bill had never been to Colorado, apparently, because in that state there are no post offices as low as Bill's.

Other places luring the tourist trade took up geographic names or relied on a feature of their particular spot. Hence the Shady Bend Tourist Camp in Grand Island, Nebraska; the Edgewood Tourist Court in Gettysburg; the Cross Roads Tourist Court in De Witt, Iowa; and the Log Cabin Tourist Camp in Cedar Rapids. Imagination was often required when conforming a name with reality—the cabins at the Log Cabin Tourist Camp were painted white and of ordinary wood-frame construction. Other names were just whimsical: the "Spen-A-Nite" Motel in Nevada, Iowa, and the Restwell Motel in Van Wert, Ohio, evoke memories of quieter days along the highway.

Of course, dozens of places took their names from the highway that passed by the door. Business people were proud to be at the edge of this great road, and more than happy for the trade it brought to the door. Lincoln View Motel and Restaurant, Lincoln Rest Tourist Camp, Lincoln Hotel, Lincoln Court, Lincoln Hi-Way Garage, Lincoln Way Cafe, Lincoln This and Lincoln That—they all stood as close as possible to the road that brought them their livelihood.

Perhaps the best-known stopping place of all along the Lincoln was Grand View Point, about seventeen miles west of Bedford, Pennsylvania, and about thirty-five miles west of Bill's Place. Here, an eastbound Lincoln Highway traveler dropped off the top of the main ridge of the Alleghenies and made a sharp turn just at a point where the view off to the east and south was indeed grand. The earliest Lincoln Highway had a widened area here and a low stone wall where motorists could pull off to view a good part of Bedford County. It wasn't long before a now-forgotten entrepreneur, intending to make stiff competition with Bill and Doc, built a gas station against the hillside

on the opposite side of the road and a viewing area at the edge, complete with souvenir stand, miniature castle parapets, and flags. Grand View Point was born. Business must have been good, for a few years later, the Grand View Point Hotel appeared, hung out over the mountainside and built right where the stand had been. The parapets were now on the hotel and a sign said, "SEE 3 STATES AND 7 COUN-TIES—MOST WONDERFUL VIEW IN THE STATE." Tourists parked at the stone wall and entered on the fourth floor of the hotel.

Only a few postcards later, the hotel had been rebuilt to become the S.S. Grand View Point—the parapets had vanished, and the square hotel had had a convincing bow and stern added, a top deck, funnels, and a crow's nest attached. A ship's wheel stood on the foredeck, life preservers were hung from the railing, and a line of nautical flags hung between flagpoles. The postcard called it "the only Steamboat in the Mountains in U.S." The nautical motif was continued inside with seascape murals on the walls of the dining room, ships' lanterns, table help outfitted like sea captains, and each chair back sporting an anchor. Rooms were referred to as staterooms, and the floors of the place were, of course, called decks.

No first-time traveler could possibly avoid being startled by this big boat, and few could avoid stopping at this ship so unexpectedly moored high in the Alleghenies.

Like every highway, the Lincoln drew the odd, the quirky, the gimcrack sort of place that had some fabulous and unique thing to offer the passing motorist. A few miles east of Medicine Bow, Wyoming, stood the "oldest building in the world": the Como Bluff Fossil Museum, "millions of years old," made of exactly 5,796 dinosaur bones weighing exactly 102,116 pounds. This place had been featured by "Ripley's Believe It or Not" and contained fossils and rocks from nearby Como Bluff, "The Dinosaur Graveyard." It called itself variously the "Museum of Famous Creations" and "The Building That Made the Fossil Famous." The address sides of the museum's post-

*Bill's Place drew travelers off the road for
gas, a cold drink and a sandwich, souvenirs,
and of course postcards.
Courtesy of Lyell Henry.*

cards spouted pretentious chaff about dinosaurs, fossils, and creation: "Truly; do these faithful backbones of life abridge the primitive pass to higher creation, reflecting a guided forward march to the essential betterments that golden this path to Destiny." Most cards had so much text that there was very little space left to tell your brother back home how the dog had been carsick for most of the trip.

Other places, while less wacky, relied on pure hucksterism. "Ye Dutch Haven" stood among the Amish along U.S. 30 near Lancaster, Pennsylvania, and boasted that it was "famous for Dutch Bar-B-Q & Gifts" and Dutch-made root beer. A sign in front of this hopelessly out of proportion windmill said "AMISH STUFF," and the place claimed that "Once You Stop You'll Never Pass."

*The Grand View Point was for many years
a popular stop along the Lincoln Highway
in Pennsylvania.
Courtesy of Lyell Henry.*

10

The Highway Takes America

Glorious, stirring sight! . . . The poetry of motion! The real way to travel! The only way to travel! Here to-day—in next week to-morrow! Villages skipped, towns and cities jumped—always somebody else's horizon!

—Mr. Toad to Rat and Mole after his first encounter with a motorcar, from Kenneth Grahame, *The Wind in the Willows*

But whatever the change beside the road, a greater change came to the road itself, and what happened to the Lincoln parallels what occurred to highways across the nation. No one—not Henry Joy, not the dreamer Carl Fisher, not the shrewd Frank Seiberling—could have predicted what would happen. No one could have known that each step of improvement—be it paving, widening, straightening, or an entirely new alignment—would so quickly summon up more and more cars and leave the job outmoded before the roadside grass began to grow back. Each time the road was changed, with a bolder and ever more costly solution, everyone thought the particular problem of traffic or safety had been licked for good, but soon it would return in a new form, in a new place with a few thousand more honking cars and trucks.

The men of the Lincoln Highway Association, like good-roads supporters and state highway officials everywhere, initially set out to pave the existing roads to allow for all-weather travel. At first, little attention was paid to engineering or traffic considerations—width, drainage, curvature, grades, sight distances, expected density of use. A paved road was good enough simply because it would allow the passage of automobiles in almost any weather. But in places where traffic was heavy from the start, like along the Lincoln through Princeton, New Jersey, great jams occurred very early, and the death and accident rates jumped as cars crashed on narrow bridges, collided head-on trying to pass where distance was insufficient, and simply went into the ditch where curves were tight and shoulders were soft. The stretch of the Lincoln that passed through Princeton was an old road that had seen very little change since the Concord coaches plied the route, and it was simply inadequate for the amount of traffic it was asked to carry.

In most places the earliest road merely got through the best it could. It followed the terrain as much as possible and sought the easiest route, the path that merely made the road passable with the least effort. Little attention was paid to directness, tightness of curves, or grades—these roads were merely wagon roads, now with tire prints from automobiles. This was the Lincoln Highway of 1915.

Beginning in the 1920s, roads everywhere were not only being paved but also widened, thickened, and straightened. Year by year engineers found better ways to speed traffic safely, but solutions seldom kept up with growing demands. Philosophies of highway engineering evolved over time and with changes in vehicles, traffic patterns, and road builders' ability to move great quantities of earth for cuts and fill.

From about 1925, when the precision of engineering was first being applied to motor roads, through the 1940s, having gentler grades was more important than having a particularly straight road. Trucks had appeared in numbers but were slow to climb steep hills, and they were a menace descending the farther side owing to inefficient brakes. Though traffic was growing, engineers of the 1930s weren't yet much concerned with sight distance for passing and safety. Mountain roads which started as crude wagon roads were rebuilt as serpentine ribbons, looping and turning back on themselves as they gained altitude gradually. They were engineered somewhat like mountain rail lines, which zigzagged back and forth across valleys and circled and tunnelled mountains to increase the length of a grade in order to reduce the necessary gradient per mile. Motorists could get vertigo on a road that was mile upon mile of tight turns and was without sufficient tangent for passing a slow truck or bus.

Year by year as traffic increased and cars grew safer and more powerful, research findings accumulated on traffic flow, safety, signage, paving techniques, bridge construction, and drainage. Federal standards decreed ever wider, straighter, more unobstructed highways. The average paved road of 1920 had been sixteen feet wide, with fences and pole lines crowding the pavement edge; by 1940, it was twenty-two to twenty-four feet, with broad ditches and obstacles set far from the reach of traffic.

By 1940, most cars had heaters and defrosters, and Packard had even introduced a primitive air conditioner. Car radios were popular, and motorists could cruise down a wider, safer, but more crowded road in comfort behind the wheel of a new Nash or Fluid Drive De Soto and listen to their favorite programs. On Saturday mornings, if reception was good, a motorist could tune in the NBC Red network and listen to "Lincoln Highway" from station WEAF in New York. After the first chords of music, announcer John McIntire would call: "Lincoln Highway! From the Main Street of America, the makers of Shinola Shoe Polish bring you a new series of living stories!"

Jack Arthur would then break into song:

> *Hi there, neighbor, going my way,*
> *East or west on the Lincoln Highway?*
> *Hi there, Yankee, give out with a great big thankee,*
> *You're in God's Country.*

Some short vignettes followed.

First man: Look at that stretch of concrete, far as you can see—bet it goes clear to the Mississippi!

Second man: The Mississippi! Why man, you see that number 30 marker? She goes all the way from the Atlantic to the Pacific! This is the Lincoln Highway!

A few bars of glorious music, then:

(Sound of engine at speed) crackling police radio: Calling state police car 24—car 24. There's been an accident on U.S. Highway 30 twelve miles east of junction with state highway 7. (Sound of siren)

Highway patrolman: Step on it Mike—that means us. The crack-up is just five miles ahead on the Lincoln Highway! (Engine speeds up)

A few more bars of stirring music, then:

Woman: There it is, darling, just beyond that clump of trees!
Man: Oh, it looks like a fine old farm.
Woman: Dad was born there. When I was a child I watched them build the Lincoln Highway!

A few strains of music, then John McIntire returns in sermonic tones. A harp plays softly in the background.

The Lincoln Highway. Yes folks, it's the Main Street of America. Trucks and jalopies and limousines, they all follow Highway 30 over the Alleghenies to Chicago, and through the fields of corn and wheat to the Rockies and the Sierras, and on west to the Pacific. Yes, this is the road that links the farms, the mines and the mills of America. And there's nothing that hasn't happened at one time or another along its 3,000 miles. The Lincoln Highway . . .

The half-hour live broadcasts used the Lincoln as a connecting element for dramas of many sorts that took place on or near the famous road. The only known recording of the show features Ethel Barrymore as Irene Mills, a distraught wife and mother, whose husband is falsely accused and convicted of murder. The program opens with her and her son westbound on the Lincoln Highway, returning home to Marshalltown, Iowa, after visiting Clark Mills in prison. The announcer says,

"A spring sunset makes the great, straight highway shine, but Irene Mills and her son, young Wally, at the wheel, see no beauty. . . ."

The story of the murder and trial is told as a series of flashbacks as Irene and Wally drive through the "corn prairie." The highway actually plays little role in the story, except that it provides a location, a backdrop for the flashbacks. Occasionally the scene returns to Wally and Irene in the car as they drive home in sadness.

Near the climax of the show, as they turn off the highway in Marshalltown, Wally says: "Mother, mother, let's not go home—it's awful here in town and I hate it. Mother, let's get back to the Lincoln Highway and head out West!"

But responsibility weighs too heavily; they continue home to reunite with Betty, the other child in the family, and resolve to do their best to free their falsely accused Clark. Betty decides to get a job; Wally offers to sell magazines. With quivering voice Irene says:

You dears, you blessed dears. We will free him together. It may take a long time, but yes, there is some way to find the truth about the murder and get Father home. . . . Oh, my children, we can't fail now when we feel this way, the three of us! We'll pledge to each other to work and plan and hope and never rest in our hearts until the man we know is innocent is free! I promise . . . I promise. . . .

The harps come up as the story ends, and the announcer returns: "Yes folks, tragedy too. But often new hopes are born along the high road of America. That was just one of the thousand stories that happen every day along the rolling stretches of the Lincoln Highway."

"Lincoln Highway" was radio's first attempt at bringing big stars to the air on Saturday mornings, heretofore a dead time in the program schedule. The show starred Ethel Barrymore, Joe E. Brown, Harry Carey, Claude Rains, Victor Moore, and Burgess Meredith and was billed as "radio's big, dramatic show in the morning." It was certainly dramatic—more melodramatic actually, in keeping with the style of the day—and also big. In the two seasons it aired, "Lincoln Highway" drew an audience of eight million listeners.

The show was no doubt popular in part because of the big stars, but it was also popular because it took place along the Lincoln Highway. By 1940, in territory east of Utah, the terms U.S. 30 and Lincoln Highway were interchangeable. When pride had anything to do with it, which was most of the time, locals called the road the Lincoln Highway. Some of the young upstarts referred to it as U.S. 30, which also worked better when giving directions to someone from out of town. The old signs were gone, the striped telephone poles had all faded, and the 1928 markers were beginning to disappear, but it was still the Lincoln Highway, "Main Street of America."

But this highway was growing invisible. As Wally and Irene drive toward home, the car rumbles smoothly and quietly; on the American highway on the eve of World War II there are no flat tires, no detours, no lack of markings, no bottomless mudholes. If adventure is to be found along this Lincoln Highway, it will be found in the lives of the travelers, not on the highway itself.

As the First World War had done twenty-six years earlier, the Second World War put a halt to all nonessential road construction and repair projects. With a critical shortage of gasoline, rubber, and construction materials, civilian travel on the nation's highways dropped to a trickle and slowed to a national speed limit of thirty-five miles per hour. Huge military convoys on a scale unimaginable to those in the army convoy of 1919 moved day and night toward shipping points on either coast. The Lincoln Highway, along with all the others in the nation, again carried the implements of warfare.

When the smoke of Berlin, Hiroshima, and Nagasaki cleared and the men and women came home in 1945, America moved quickly to get on with life. The highways had been ignored and had suffered under excessive wartime loads; now the nation began to remake them in grand fashion. Postwar America was no longer satisfied with the old roads; they needed to be wider, straighter, and faster to carry the travelers and commerce of a nation impatient to get on with life after the world's most horrible war. The Lincoln Highway and highways across the country were rebuilt to speed the nation.

Several parts of the Lincoln were rebuilt to four lanes in the early fifties. In California, New Jersey, Pennsylvania—even in Wyoming on either side of Cheyenne—the Lincoln was widened to carry a greater burden of peacetime commerce. These new roads greatly enhanced safety and the flow of traffic, and solved the problem for a time, but it seemed they brought more cars, buses, and trucks to the highway and funneled them into the centers of towns and cities, moving the difficulty to another place.

Soon highway departments started building bypasses around especially congested areas in cities and towns. The first of them diverted traffic a block or two off of Main Street, usually through residential neighborhoods, but traffic soon grew too heavy and entirely new routes were constructed, routes that skirted the towns altogether. In and around some medium and even fairly small communities, the Lincoln Highway, like many of the early automobile roads in the country, was improved, moved to a new path, widened, and moved again to a route outside the town limits. By the 1980s what had once been the Lincoln Highway had been moved in some places as many as four times.

In the vast rural reaches of the country, the Lincoln Highway and all the others were built over again and again, occasionally on the original alignment, sometimes moved a great distance; sometimes piecemeal, a curve or hill at a time. At other times, one long stretch of road, thirty or more miles, would fall under the bulldozer's blade to come out unrecognizable from the original.

Then came the interstate. Where other highways had been wide, this one was wider. Where others had been straight, this one went like an arrow. Where others had gouged and scraped the landscape, this one remade the face of the land wherever it went. Where others had been big, this one was enormous—42,500 miles of four-lane, limited-access, high-speed roadway to link all the major cities of the nation. This sort of highway was as new an idea as the Lincoln had been in 1913.

At midcentury President Dwight D. Eisenhower—the same man who had struggled across the continent via the Lincoln Highway in 1919 with the army—signed what is known as the Interstate Highway Act of 1956. Eisenhower hadn't forgotten his long convoy trip to the West Coast. He also remembered Germany's autobahns and understood how they had helped speed Hitler's military machine. As they had been since the 1920s, America's highways were to be built in the name of national defense. The Interstate Highway Act authorized construction of a highway network that promised to hurry the nation's commerce and military, unclog the streets of the cities, reduce high-speed collisions by separating opposing traffic, and greatly reduce driving time by eliminating stoplights, sharp curves, intersections, and no-passing zones.

This was a new strategy for a new time. By 1956 the travel patterns of the nation had changed; Emily Post had declared her dislike for slow, dirty trains forty years earlier; now everyone had abandoned the rails for the privacy, comfort, and convenience of the family auto. Vehicles of all sorts—cars, light and heavy trucks, motorcycles, buses, mobile homes, campers—had appeared from everywhere. Speeds had greatly risen and heavy trucks could better hold their own. Ever straighter, wider, and faster roads were in order here, highways that interfered the very least with traffic movement. New earth-moving and road-construction equipment made the ideal nearly attainable. Great machines were set to work moving huge volumes of earth that

would have been impossible for the highway engineer of 1925. They gouged great chunks of mountain, hillside, farm, and river bottom to build a road gentler in grade and curve than earlier designs. Mobile concrete plants could easily pour a foot-thick, two-lane ribbon of cement a mile long every day. Could Arthur Pardington have imagined that when he shipped those barrels of cement to De Kalb County, Illinois, for the first seedling mile?

Now, a new mountain-pass road might follow a more or less straight line diagonally along and up a mountain ridge and over the top, as Interstate 80 does today over Donner Pass. To cross a valley, the new four-lane interstate would be built straight across; the valley sides would be cut and the fill dumped in the middle, replacing the older road that angled this way and that, picking and choosing its way down one side and up the other. Trucks could now drop over the edge and build up momentum on the downgrade to use for the climb back out, rather than braking their way through curve after curve on the old downgrade and grinding their slow way up the farther side.

The interstate is really the fourth kind of automobile highway to grace the landscape. While George R. Stewart gave us four kinds of automobile roads based on their origins—those that began as foot trails, early turnpikes, routes along the railroad, or routes built new for the auto—these other four types detail the evolution of roads after the auto and truck became the primary vehicles to use them. Since the advent of the motor car, roads have undergone constant reconstruction and route change, and this evolution can be clumped into rough stages that yielded four kinds of highways. A look at a place near Grand Junction, Iowa, will serve to illustrate all four.

Just east of this town, on the north side of the Chicago and North Western Railway, an old and narrow concrete bridge crosses a small creek. At either end of the span is cornfield; all other signs of the earliest route of the Lincoln Highway have vanished under the plow.

On the south side of the tracks is a newer bridge over the same stream that carries a newer but also bypassed route of the Lincoln directly into Grand Junction. A few steps farther south is the modern path of U.S. 30, making a wide sweep and climbing an embankment to cross the tracks and bypass the town altogether. But even that up-to-date road doesn't carry the cross-country traffic: trucks, cars, buses, and campers with out-of-state plates pass thirty-five miles to the south on Interstate 80.

These four roads—or fragments of roads—illustrate a chronology of road development since 1900. These characteristic highways can be found along the route of the Lincoln and many of the other roads in the country. Sometimes this evolution produced separate routes as it did near Grand Junction; sometimes the changes occurred all on the same path.

First is the earliest path used by the automobile: usually the primitive road that saw the first gas buggies and was first marked as the Lincoln Highway. When it was improved, the work was accomplished by small-time methods, most notably with shovels and horse-drawn equipment. In some places it was paved, and when it was, this path had corners instead of curves and went right down Main Street in every town it passed. The gas stations had peaked roofs and high, glass-topped pumps, the cars were Overlands and Model T Fords, and the landscape pressed close to the road.

The second stage usually came after 1920, when highway engineers realized that piecemeal upgrading was not enough, that major changes in grade, alignment, and bridges were needed in order to accommodate heavier traffic. This is the "mezzohighway" of the 1920s, 1930s, and 1940s, when extensive pavement came and our familiar two-lane federal roads emerged. It is the highway of the diner and cafe, the road of Burma-Shave signs, narrow bridges, the Chrysler Airflow, and truck drivers who wore eight-sided caps. This is the highway of neon and mom-and-pop tourist courts, scenic overlooks, and gas stations with all-electric pumps. Along this second-generation highway the great boom in roadside America took place.

When the Lincoln Highway left the earliest zigzag route across Nebraska for the route along the Union Pacific tracks, it became a second-generation highway. While U.S. 30 still follows the tracks across Nebraska, in many other places the second route was in turn bypassed by a third, leaving the mezzohighway as a back road and more or less frozen in time.

This third stage is the bypass: the two- or four-lane edge-of-town route that took the federal highway off of Main Street. In many ways it is the embryonic interstate; it is four lanes in places, built with wide shoulders and ample sight distances. But unlike the interstate, bypasses did not always separate opposing traffic. These roads have intersections and stoplights, left-turn lanes, stop-and-go traffic. Along this roadside rises the suburban strip with shopping centers, giant auto dealers, frontage roads, and housing developments.

The fourth stage is, of course, the interstate. Taking a cue from the urban bypass, the interstate headed for open country, for an entirely new route whenever it found an advantage. This new sort of path is not bound by the old; its view is long. Where the mezzohighway joined two main and distant points by connecting all the intermediate cities and towns like beads on a string, the interstate might connect the same distant places with a new straight line having little regard for other places along the way.

Like the Lincoln Highway of 1913, the interstate system of 1956 was a new sort of line on the American landscape. It demarked a change in what roads could be, changed the way Americans traveled, and changed the way they looked at the land.

Interstate 80 became the highway the Lincoln was meant to be. It became the main line across the middle of the country, and at last there was a highway with a single number that connected New York

City with San Francisco. U.S. 30 and 40 withered in the shadow of the new red, white, and blue shield, and the Lincoln Highway grew even more faint. Interstate 80 is 2,906 official miles of unbroken, uninterrupted four-lane pavement—no street corners, no narrow bridges, no rural mailboxes, no Burma-Shave signs. Broad and straight, it begins at Teaneck, New Jersey, where it splits off from Interstate 95 just a few miles west of the George Washington Bridge over the Hudson River, and ends just off the San Francisco Bay Bridge in the middle of San Francisco, near First Street. And although it begins and ends a little short of either ocean, L. B. Miller and C. I. Hansen could now make their 1925 trip direct and nonstop from New York to San Francisco instead of from Jersey City to Oakland. They could do it in just two days—including stops for gas and food—and still never exceed the speed limit.

Henry Joy would have approved of the route; quite possibly he would have selected the same or a similar path for the Lincoln had he not been confined to the existing roads and road-building technology of 1913. Where Joy had to compromise on his route, Interstate 80 pushes directly through. Interstate 80 avoids the Lincoln's angle south to Philadelphia before tackling the Alleghenies; it starts boldly west from New York and threads the Delaware Water Gap in Pennsylvania and takes the mountains head-on. Following the example set by the Lincoln, Interstate 80 never enters the congested hub of Chicago, and later it follows a good part of the Platte River valley across Nebraska. And from Salt Lake City, it goes to Wendover across the salt flats— a place tamed by modern construction methods—and follows the Humboldt River and California Trail route to Reno, leaving the Lincoln far to the south in the desert of Utah and the summits of Nevada. Interstate 80 adds one more vote to the good choice made by many California emigrants, the Central Pacific Railroad, and the Lincoln Highway: it crosses Donner Pass, though at a point two miles north of the main cleft.

Even if we add, say, twelve miles to get from Times Square to Teaneck and seven miles to take us from the Bay Bridge up the hill and across the peninsula to Lincoln Park, overlooking the Pacific, we still have a route much shorter than the Lincoln—2,925 miles for Interstate 80 compared with 3,185 for the Lincoln in 1931, a difference of 260 miles. This is the highway of Henry Joy's dreams.

Although it is efficient, fast, and safe, the interstate highway system has become the ultimate bypass. Wherever it goes, it controls the landscape, then becomes the landscape itself. The oil-stained band of cement and asphalt has made every part of the country hours closer to every other, but it has also made it easy to pass it all without seeing any of it. The world exists somewhere beyond the exits, which are now usually referred to by number instead of by the places they go. The signs might as well say, "Exit 26—gas, food, and lodging—assorted small towns." The interstate has created the environment, the fertile soil, that drew America to the broad pavement. It is an insular environment where we steer vehicles designed to smooth every bump, exclude sound, scent, and dirt. We fill them with equalized stereo sound from somewhere else and arrive at our destination fresh, but can we say anything about where we've been?

Certainly the interstates and the roads that spring from them have made auto travel safer and faster for millions of motorists and truckers. There are many highway patrol photographs from the 1950s of grinding high-speed accidents on two-lane highways where vehicles came together with great impact, often in heavy traffic when an impatient or drunk driver pulled out of a line of cars to pass and suddenly met an oncoming vehicle. Unrecognizable smashed metal, a puddle of oil—or is it blood? While plenty of horrible accidents still occur, the frequency is much lower with median strips separating opposing traffic, wide, unobstructed ditches and shoulders, and guarded abutments.

The interstate system speeds the commerce of the nation and allows a motorist a range of travel unimagined fifty years ago. Millions of travelers and tons of goods arrive sooner and cheaper than they ever could have via earlier highways.

But Americans driving from New York City to San Francisco today no longer see or interact with the same landscape that so captivated automobile travelers early this century. In the past, travelers on the Lincoln Highway traveled deep amid the cultural and natural landscape of the nation—down the shaded streets of midwestern towns, close past farmsteads where one could see what kind of chickens the farm wife raised, through deep forest, close enough to the salt desert that the eyes burned—the grand montage of detail that adds up to the impression of a trip, and of a country. Today you can cross the entire state of Wyoming and never smell sagebrush. The Great Salt Lake Desert that haunted travelers for 120 years is a remote abstraction when seen from the wide ribbons of Interstate 80 amid campers and throbbing trucks; Main Streets are nostalgic reminders of the "olden days" when towns pass by in the distance, a mile or two from the interstate. The places to eat and to stay and to buy gas along the main highways are no longer manifestations of a particular place—the rural South, the urban East—but emblems only of the highway, a great franchised monoculture that extends from sea to sea.

11

The Faint Line

Out across the country, usually just out of earshot of the restless trucks on Interstate 80, I have looked for the routes of the Lincoln Highway. In most places there are two or three, and all together on a map they look like a piece of old braided rope. This cord of routes is raveled and frayed in places where there have been several paths, and it is close and thick in places where the route has always been the same. At several points across the country, the rope is broken entirely, with whole sections missing, gone beneath superhighways and suburbs. Today the strands lie loose across the landscape, just under the surface, overlooked by most people and overshadowed by newer lines of travel.

Out along this faint trail are still a few of Gael Hoag's concrete Lincoln Highway markers. Of the original three thousand, perhaps only ten or a dozen still stand along the highway where the Boy Scouts carefully set them in 1928. Most have fallen to road construction projects and roadside development, some to vandalism, and a few must still be out there, lost to shrubbery and undiscovered, right where Hoag noted they were to be placed when he drove the highway for the last time and made cryptic pencil marks in his field notebook. From time to time someone finds one in a barn, or part of one in a ditch. A man in California told me that somewhere in the Sierras there was a rubble pile, a landfill where dozens had once been tossed and lay at the bottom. A fine one stood solid and serene between the highway and the Union Pacific tracks just east of Potter, Nebraska, until recently when it was removed to regrade an intersection.

Many have been saved or recovered, to be placed in parks, a museum or two, a garden. At least one was moved from its original spot on the boulevard to a flowerbed by the front door of a house. Its protector worried that a careless driver would run over it and that it would be lost for good.

But out along that 3,300-mile fragmented road are a few. My wife, Carol, found the first one we'd ever seen, hidden in the grass near an intersection at Tuscarora Summit, just across the highway from where "Doc" Seylars Rest House had been. She turned to look for oncoming traffic and just caught it with her eye. Later we found another standing amid the mountain laurel near Fayetteville, Pennsylvania; one at curbside on the east edge of Lisbon, Iowa; a handful in towns in the West, perhaps in original spots, perhaps moved.

The 1928 markers are the symbolic icons for the Lincoln Highway of Henry Joy's time and for the Lincoln Highway of our own. They were originally placed to mark this road forever as a continent-wide monument to an American idealist and to mark the path of an ideal highway. They were placed with great hope that this highway would be remembered even after a system of numbers erased its official status. Whether the directors and officers of the Lincoln Highway Association thought the markers would actually keep the highway alive is not known, but their idealism is expressed by even the few that remain.

The markers are emblems for the Lincoln Highway today. They stand as fragments, like pot shards or ceremonial beads or other objects of a lost culture, a culture overrun by a faster race. Fugitive, scarce, secretive, and like pieces of an ancient water jar, there are too few to make an entire vessel. And though they are the compass points to the old Lincoln Highway, the path of the road cannot be divined from them. There are too few; they are too scattered to form any sort of line across a continent. We have to use other totems, other artifacts that establish the route and help bring fragments of the once famous highway to view.

The 1928 markers were once imbued with great meaning, but the people who revered them are gone, leaving only their symbols, the objects that still glow somehow with light of meaning. I always come upon them unexpectedly and feel like the archaeologist reaching quickly for that glittering bit of glaze in a sifting screen full of blind

soil. They were always new and yet always familiar. Whenever I've run across one, I've thought of Gael Hoag sitting in the driver's seat of the Packard, motor idling impatiently, making a note in his field book for the marker's eventual location. Why yes, I can see why he put it there.

In the cities that the Lincoln touched, almost all trace of the Lincoln has disappeared. In Pittsburgh, Sacramento, Salt Lake—even much smaller places like Laramie or Cedar Rapids—the paths of the road are ordinary city streets with nothing to distinguish them. Often the only clue will be an old Lincoln Court Motel or Lincoln Cafe, so remodeled and rebuilt as to belie its past. In such places the city dominates the highway into submission. If the Lincoln is to be found at all, it will be in the countryside.

The older the route, the less likely there will be anything left to find. The oldest have been subjected longest to the onslaughts of time and change. Certainly the old guidebooks have helped—the *Blue Books,* the *Complete Official Road Guide of the Lincoln Highway,* and others—but in some places, even rural areas, the roads have changed such that the early routes are entirely gone. Old road maps were very useful, especially a progression of them over time of the same state that shows the evolution of the route. Sometimes I can tell where the cartographer fudged; I can see he didn't have any idea where the road went in a certain place; his lack of information caused him to draw it in as a vague wavy line. To make it too straight would call attention to it. Topographic maps helped as well because they may show very old roads of local use or roads that have now been fenced off. Local inquiry was sometimes helpful, but usually other sources were more accurate, more comprehensive.

The sharpest tools are eyes trained to find old highways. A scramble down the stream bank under the new bridge reveals the abutments of the old. A newer highway comes under suspicion when cracks appear about two feet from either edge, revealing widening strips added to once-narrow pavement. An old advertising sign standing in the sagebrush a half mile from the new highway, a grove of trees with an old gas station in front—was this a tourist camp? A faded sign painted on a barn, an eroded embankment tell me of travel years ago. Sometimes it's the way the road feels under the car, sometimes the way it angles away from the main road; sometimes it's a bit of old guardrail, a narrow concrete bridge, a Lincoln Highway marker. And it's almost always Main Street.

A traveler following old routes of the Lincoln Highway today can no longer mail a letter from Bill's in Pennsylvania, home of the "highest and smallest post office"; the place has disappeared as if it had never been. The Como Bluff Fossil Museum in Wyoming is closed, but the building is still there. A building that many millions of years old would be hard to tear down. You can't get a steak dinner at the S.S. Grand View Point Hotel because it is closed too. There are no little cabins for rent at the Kampus Camp in Ames; there is no place to dance in Bosler, Wyoming, anymore, and they've been out of gas for years at the Summit Garage in Altamont, California.

The S.S. Grand View Point is still there, poised on the edge of the mountain above three states and seven counties, but it is now called Noah's Ark, and an owner has of late taken to covering parts of it with cedar shingles to make it look more like the ship from the great flood. The job is half finished and looks like it never will be completed. The structure stands forlorn with its big telescope drooping uselessly on the front deck. A novel, once memorable and quite beautiful tourist attraction slowly decays into an eyesore.

There must be people alive today who as children visited this place with their parents and recall the white ship in the mountains with that special memory of childhood, the intense, indescribable flavor of something so wonderful, so unexpected, so memorable to a young

mind. Does he or she still cherish a postcard or curio from Grand View Point? A faded felt pennant? "Grand View Point Hotel—See 3 States and 7 Counties." How could Seven Flags over Anywhere possibly compare?

Not many stop at places like the S.S. Grand View Point Hotel anymore. We'd worry that the food might be bad or the beds unclean, and a mere Pennsylvania view is hardly enough to slow a traveler on the way to the Grand Canyon.

For the most part, today's roadside entertainments, meals, and lodging are institutionalized, franchised, sanitized, and market tested. We seem to be unwilling to take a chance when we travel. We instead seek security and blandness in chain fast food, chain lodging, and programmed entertainment. A nationwide hotel chain boasts "no surprises," and we drive all day from one to the next and find the same painting on the wall that we did the night before.

America savors speed, the shortest time between origin and destination. Driving distance is measured in hours, never miles. We gauge the quality of the trip based on how much of it will be four-lane—all interstate is a special bonus. Travelers seldom look to either side while en route; the view is always forward, through the windshield toward a motel reservation, business meeting, national park, or grandmother's house. The impulse of haste marks us all on the interstate. We pull out to pass a truck going a mile per hour under the limit and find ourselves barreling along at flight speed with the grills of a line of impatient vehicles in the mirror. Soon we take up the chase and are swept along like flotsam headed for the rapids

But a curious thing happened. Once off of the main roads, once slowed down to subsonic speeds, I found that the fabric of travels could be very similar to those of Bellamy Partridge or Emily Post or Beth O'Shea. My wife—who accompanied me on many trips on the Lincoln—and I found it was easy to travel slowly, with our eyes open.

We clearly had some advantages over earlier motorists. They usually traveled slowly because they had to. We never got stuck and never had to worry about the effect of any cloudburst on our progress. We had modern maps and could step onto the main roads and travel quickly when we wanted; early motorists had only our back roads. For the most part we drove a better car than they did, and even the bypassed roads of today are in much better shape than the paths they had to drive on.

The pace and impression of travel, though, were similar. We traveled slowly to be absorbed by the landscape, to look for easily missed roads, and because the roads were sometimes rough. We stopped often to look, to photograph, to eat, to rest, to ask directions, and just to enjoy the view. One day on the road we went a total of fifteen miles. We pitched our tent almost within sight of where we'd camped the night before.

Even when I traveled alone, strangers showed me many kindnesses. I was offered meals and a bed on several occasions by people I'd only just met. I couldn't always or even often drive the old routes of the Lincoln Highway for mile after mile, narrow between close fences, looking for red, white, and blue telephone poles, reading the *Blue Book* for every clue, but I did it enough to get a feeling for what it must have been like.

In order to mimic Thornton Round and his tire-repairing antics, I even changed a couple of tires. They were damaged on an abandoned stretch of pavement near Evanston, Wyoming, where a contractor had dug a pipeline across the old highway and left the ditch partially filled. The ranchers who traveled the road undoubtedly knew where the break was, but at the crest of a hill, I left black skid marks leading into the deep hole. Like out-of-town travelers in past years, I found difficulty where locals knew the ropes. I didn't spend my evenings around the campfire patching tires and tubes as Round had—I bought a pair of new tires.

There were no true tourist camps for us to spend the night in, but we camped in pastures with permission and almost anywhere we pleased in the dry, open country of the West. We met no Road Liars as Frederic Van de Water had, but in private campgrounds—not the big commercial affairs, but the little mom-and-pop places, off away from the big RVs in the back areas for the tents—we met a cross section of wandering America.

I sometimes wondered how much of the Lincoln Highway Henry Joy or Harry Ostermann or Gael Hoag would find familiar today. Not much, I'd guess, around the cities across the country. In the more rural areas and smaller communities, they might be more at home. But even here the road has been remade so thoroughly and so many times that probably only the names, the general lay of the land, and landmarks—like Turtle Creek near Pittsburgh, or Robinson Summit in Nevada—would ring clear and be recalled. But even a landmark so distinct as the Mississippi River crossing between Fulton, Illinois, and Clinton, Iowa, would send them scrambling about for clues; the old bridge is gone and the levee made over so completely in recent times that only street names and tiny landmarks remain to signal where the rickety high bridge once launched itself across the river.

They wouldn't be lost exactly; they were too observant to get befuddled and misplaced, but much of the time they would be traveling in a foreign landscape. In much of the West the dry, unchanging topography would be reassuringly familiar, but even here the paths of travel have moved time and again such that the old ways would be hard to recall.

The earliest path that Joy and his people knew so well is the thinnest strand and is the most likely to have disappeared. It may be abandoned and lost under sagebrush, volunteer forest, or cropland, or most likely torn up and buried by America's constant remaking of the landscape. In east suburban Fort Wayne, Indiana, a few hundred feet of early concrete cuts across the parking lot of a trucking company.

Beyond in either direction it disappears beneath new highways and industrial construction. Across much of Wyoming, the early trail Alice Ramsey followed in her Maxwell has vanished under the wheels and treads of four-wheel-drive vehicles and dozers laying pipelines. It was for many years a very informal path, following the terrain like the wagon trail it was, changing its course when circumstances dictated, never leaving much mark on the land. But just where did Ramsey cross those arroyos, just where did the ruts lead her? From Interstate 80 today—the only path across most of southern Wyoming—little can be seen of Ramsey's path.

In a few precious places the oldest road still awaits the flying wheel in nearly original condition. In Nebraska, near Elkhorn, the original paving—narrow, bumpy brick laid in 1919 and 1920—stands open and drivable just as it did when Harry Ostermann made his last trip west. Now preserved by a local-history group, it is a true living monument to the early highway. Another sliver of the oldest road stands near Kingston, New Jersey, another bit just south of Ligonier, Indiana, and tiny fragments stick out here and there in California, Ohio, Iowa, and many places the Lincoln Highway went.

But west of Salt Lake City, in Utah and eastern Nevada, out toward Fish Springs and Tippett on the salt desert and in the mountains and valleys beyond, the fragment is better than one hundred miles long across territory that has changed little since the early days on the Lincoln. Here any one of the old Lincoln Highway men would be right at home.

Near the west edge of the Great Salt Lake Desert, along the road between Callao and Gold Hill, Utah, I made an overnight camp a few summers ago. On this trip I was traveling alone. The road had begun its climb out of the salt desert basin into the Fish Creek Range, and from where I pitched my tent, I could look back across the salt to the mountains and rocks I had passed many hours earlier. It was near sunset when I had found my campsite among the sagebrush, and as the light faded, the wind died away.

I had been driving the earliest route, the path along the southern edge of the salt desert through Fish Springs and Callao. The Goodyear Cutoff was now inaccessible in the middle of the Wendover Bombing Range, a pockmarked no-man's-land of craters and targets for aerial gunnery.

I knew that this was as close as I could get to the authentic early highway, the road of Henry Joy and Gael Hoag. This path across the desert had certainly been improved some in the years since the Lincoln Highway Association Packards had come this way, but the landscape and the experience of travel across this remote place had not changed. It was as dry and uninhabited as it had been in 1915. The road was a little better—the Thomas mudhole had been bypassed long ago, and where there had been stream fords there were now culverts. The road was graded occasionally and had an embankment in places, but it was still slow, dusty, and hot. I imagined that my non-air-conditioned vehicle with 90,000-plus miles on the odometer compared about equally with an open Packard in terms of comfort and reliability. Like early travelers here, I carried a few extra parts, tools, twenty gallons of water, and a handful of guidebooks. Unlike early motorists, I carried an iced cooler and cooked over a camp stove instead of a campfire.

In some ways this path between Salt Lake City and Ely was more remote than it had been when the early autoists came west for the Panama-Pacific Exposition. Old man Thomas was long gone, his hut and bunkhouse torn down, and many others had left the scattered ranches and settlements. At the time when this was the Lincoln, a traveler could buy gas and find lodging at a few places along the way; now it was a long dry spell between Tooele and McGill, some 220 miles.

As my evening meal simmered on the camp stove, I looked east toward the desert and could faintly see the Goodyear Cutoff, a thin, straight line coming around the north end of Granite Peak, crossing the salt for some eighteen miles; and I imagined what it would have been like to be stuck in the middle, like a fly on immense white flypaper, and to be there as night fell. The salt desert is pool-table flat and holds almost no vegetation. A rancher at Callao once told me about how the flashing headlights of mired autos had sometimes been seen in Callao, twenty to thirty miles southwest of the Goodyear Cutoff.

Around the south route through Fish Springs, the old road twists back and forth across the new. I had stopped often to walk a few steps or a few hundred yards to one side or the other to find the faint tracks of Mark Twain's alkali-covered stage coach, Alice Ramsey's Maxwell, and Henry Joy's Packard. The 1916 Lincoln Highway guide says that between Callao and Ibapah "care should be exercised as washes are liable to be encountered, which if taken at too great speed, are liable to break a spring." Indeed. In a short walk along the old path, I uncovered two broken springs left long ago.

As nighttime set in on my tiny camp in the desert, I ate my meal and noticed that from where I sat I could see only one light, far across the desert. But despite its remoteness I somehow had a funny compulsion to leave the coffee pot on the stove to warm after I'd poured myself a cup. It was as if I was expecting visitors. The road below my camp was empty, as it had been all day as I drove west, and the desert was without sound except for my own heartbeat. But in my mind's eye I imagined company.

The pony expressman wouldn't stop—he has a schedule to make— nor would the overland stagecoach. But Henry Joy might. As I sat and looked at the stars and the black desert, I imagined a pair of close-set headlights appearing in the distance from Callao, and a heavy Packard touring car pulling up and Joy stepping out, lifting his goggles and pulling off his gloves.

But the desert remained silent; not a car nor horse passed while I sat, nor through the deep night as I slept.

Because the desert section was bypassed so early, it retains the simplicity of the earliest highway, but it missed all the roadside development that came to the mezzohighway. Iowa, Illinois, and Nebraska may be the happy hunting ground for the intermediate Lincoln Highway. In Iowa, routes older than present-day U.S. 30 can carry the traveler across half the state. They are now county roads that carry light traffic and run right through the centers of towns like Lisbon, Belle Plaine, State Center, and Scranton. There is a fine, uninterrupted forty-mile piece between Marshalltown and Ames.

In Illinois between Geneva and Sterling, some eighty miles of Lincoln Highway were left as secondary highway when U.S. 30 was moved wholesale to a direct and straight path a few miles to the south. This route is now Illinois 38 and Illinois 2. In Nebraska, U.S. 30 still runs through the centers of forty-three towns; highway development skipped the bypass stage and jumped right to Interstate 80 running parallel and a mile or two south across much of the state.

In these places the old road has reverted to a sort of easy-paced, county-to-county highway that bears a certain resemblance to the highway of years ago. That resemblance is somewhat architectural, for these are the routes where the artifacts of our automobile culture lie, but it is also a cultural resemblance, a similarity between the community of the highway there today and that of years ago.

When the main-line traffic moved to the bypasses or the interstates, these old routes stalled, and some began a reverse evolution of sorts. Where they had been accelerating toward that commercial highway we all know so well, when the long-haul traffic left, many began to regress in a way, to fall back to something like what they had been earlier. Much of the architecture did continue to change, and the roadscape has continued to evolve, but the role of these paths in a mobile culture may have reverted to something travelers in the 1920s might recognize.

When the out-of-state license plates moved to Interstate 80 in order to get across Nebraska, U.S. 30—though still a federal highway—fell to local-road status. Grain trucks, salesmen, farmer's pickups with livestock racks, and cars with Nebraska plates became the predominant types. The little downtown cafe or truck stop on the edge of town ceased being a place crowded by eighteen-wheelers and impatient travelers—the new one on the interstate took care of them. The old place reverted to more of a local spot, a place where the rural postmen stopped for coffee and the waitress knew almost everybody by a first name. Even if lumberyard paneling and formica hid the old ambiance, it was still a place that in a way might be familiar to earlier travelers. The social milieu might echo what earlier travelers found so appealing about places far from home. It ceased being a place only of the highway, only of someplace else, and became a mix of local and regional flavor.

Along this road we found the fellow who sees the license plate and says, "You're a long ways from home!" Along this road we stopped in the Corner Cafe in North Bend, Nebraska, for breakfast and the waitress remembered us from the last time we had stopped—a year earlier. Along here are not only the architectural relics of the Lincoln Highway but the spirit of the old road. Most of the people older than forty in gas stations, cafes, and grocery stores remember the road and have their own stories to tell about the day it was paved, a blizzard, an accident, or the time they drove it clear to one coast or the other.

This is the often narrow road that passes box gas stations and cafes with curtains in the windows, and crosses steel-truss bridges. In a place or two it may even split a farm: the house on one side of the road, the farm buildings on the other. It is an extension of Main Street wherever it runs, and even though it doesn't run down Main Street in Belle Plaine, Iowa, that town represents what the Lincoln Highway was in its prime.

Belle Plaine began as a railroad town along the Chicago and North Western but quickly took to the highway when the Lincoln came through in 1913. Though convenience stores, car dealers, and supermarkets have sprung up along the old highway through town, a slow drive and a few stops reveal much about the Lincoln in the days prior to 1936, when it moved to a route several miles north. There are several surviving places from the salad days along the Lincoln—the Maid-Rite Cafe, the Lincoln Cafe, E. L. Sankot's Garage, the Iowa Hotel, and the Herring Hotel. In the 1917 *Automobile Blue Book,* the Herring Hotel billed itself as the "Swellest Little Hotel on the Lincoln Highway." Travelers were invited to "stop and knock off some of the mud or dust, as the case may be, and get a Souvenir Postal Card anyway, whether it is meal time or not. You are welcome. Rooms with baths. Cordially yours, Will P. Herring & Son."

Not everyone thought the Herring so swell. Dallas Sharp stayed here in 1927 and wrote, "I shaved in cold water and thought of travelers in swift, warm sleeping cars." Today the Herring Hotel is the Graham House, a home for apartment dwellers rather than tourists.

It is hard to be a stranger for very long at the Maid-Rite Cafe. There are no booths or tables, and the stools around the two horseshoe-shaped counters make it seem like everyone is at one big table. The place was last remodeled sometime in the 1920s, but nobody remembers just when. Farmers and businessmen come in for coffee or breakfast, talk about crops, weather, neighbors. A question about the Lincoln Highway touches off a debate about when the road was first paved, who first operated the tourist camp, and just when the state will get around to fixing the road west of town.

A couple blocks west is Sankot's Garage, where Roy Sankot sits on an old school-bus seat and waits for repair business or someone to talk with. Lining the walls are shelves full of parts for cars that haven't been made in years. Roy's family has operated a garage and repair business here since 1914.

On the west edge of Belle Plaine, across from the railroad yard, is a latter-day landmark and shrine to the Lincoln Highway. George Pres-

The Maid-Rite Cafe, Belle Plaine, Iowa.
Photograph by the author.

ton's gas station is known to travelers around the country who stop to have a bottle of pop, look over the place, and talk with George. Nailed to the outside walls of this early-twentieth-century wood structure are vintage metal advertising signs hawking gas, oil, tires, and the Lincoln Highway. Old cars, steam tractors, and gas pumps stand about, and there is an old Lincoln Highway marker at the corner. Inside are maps, calendars, news clippings, postcards, and relics from the days of eleven-cent gas and a $1.65 brake job.

The main highway bypassed Belle Plaine for a straighter path, but in some places the Lincoln Highway never moved. There are several places where U.S. 30 or another major highway today stands on the exact path Henry Joy laid out in 1913, and in the case of the Lancaster Pike, directly on top of a road with a history at least 120 years older than the Lincoln. In a process akin to sedimentation, layer upon layer of improvement was applied over the original path. The road has been straightened and widened, and in places U.S. 30 veers off to a newer bypass, but for the most part the semis run today where the Conestogas once did. The road evolved in gradual steps from a macadam road to a broad federal highway—where the roadside landscape was pushed farther and farther back and the pavement was laid ever deeper and wider to make a modern path.

The road is busy, surprisingly so considering that the Pennsylvania Turnpike runs parallel a dozen miles north. Heavy trucks mix it up with all manner of other traffic from farm equipment to campers. A part of it is three-lane, a dangerous idea from the start owing to the calamitous mix of left-turning traffic with high-speed passing traffic from the opposite direction. On the back of a pickup I saw a bumper sticker, "I survived Highway 30," and a roadside sign warned, "Caution: new traffic patterns ahead." Would the cars be doing square dances over the next rise?

The origins of this road as an early path to the West are still apparent despite wholesale development along the highway; a few old inns and Georgian homes stand with their steps at the edge of the widened pavement. The road turns gently here and there, giving clues that it was not originally laid out with a surveyor's transit; a modern highway engineer transported back to 1790 would run it with laser equipment straight as an arrow for miles.

I imagined a core sample from the center of the road revealing a foot of new concrete or asphalt on top of older thin paving from the thirties, and below that, perhaps paving brick from the early days of the Lincoln. Beneath would be gravel and perhaps decayed planks from the middle and late nineteenth century, when the railroad reigned and the turnpike fell to disrepair. Under it all, just above the untouched Pennsylvania soil, would be the close-packed macadam of the Lancaster Pike.

Not far from Paradise I spied an old Lancaster Pike marker in the front yard of a modern home. It was set back a little from the road and watched over in the same way that Lincoln Highway markers have been. Lancaster Pike markers are 130 years older than Lincoln Highway markers and much more scarce. This one gives the distance both to Philadelphia and Lancaster: "51 M. to P. 11 to L." How many westbound teamsters spied this marker and weighed their remaining endurance, the strength of their oxen, and the hours of daylight left, and wondered how close they'd be by nightfall? The old stone is pitted and worn, and looks out of place as an ornament in a manicured lawn, but someone has sought to preserve it, and the old milepost stands waiting to remind observant travelers that this road is older than most. But few speeding past today see it. The teamsters are at it yet: I waited for a dozen trucks to pass before I could recross the road to my car.

We roll up a good many miles on the interstate system, Carol and I, especially on Interstate 80, which runs only a few miles from our

Roy Sankot, Belle Plaine, Iowa.
Photograph by the author.

George Preston and his station along the
Lincoln Highway, Belle Plaine, Iowa.
Photograph by the author.

Tippett, Nevada, Antelope Valley.

A house, a barn, a few sheep and a small parcel of cultivated land, is all that Tippett appeared to be to the chance passer by. A short visit, however, soon convinced us that there was more to the place than we looked for. We found a post office, a warehouse well stocked with merchandise, and an interesting lot of people.
—Hugo Alois Taussig, 1909
  Retracing the Pioneers from West to East in an Automobile

Abandoned Lincoln Highway near Robinson Summit, Nevada.

We decided to drive off into the night getting out the cots when sleep overtook me. We did, but such a road. You can hardly lose your way, but that night we went right over sage brush and everything. It was cold too. We got stuck in the rocks once, high mountains around us.

About midnight I pulled up, Alice and Dan sleeping on cots, and the rest of us in the car. We had no idea where we were, but we slept like the dead.

To make it all the more weird I thought once the Ford was on fire, but it was only a shooting star. Alice says the coyotes yelled, but the rest of us didn't hear a sound.

No other state gives the sky impression that Nevada does. Being particularly clear the stars seem very close, and they shoot a great deal with lots of light.

—D. Norman Longaker, 1923

*Brown's Hotel and Eureka County
Courthouse, Eureka, Nevada.*

*It was nearly nine o'clock when we came
into Eureka, and drew up at the dim lights
of Brown's Hotel. Brown's Hotel seemed to
be mostly a bar room and lounging place; at
least that was the impression made upon me
by the glimpse I caught of the lighted room
downstairs as I stood on the wooden porch.
But we were shown upstairs to a very comfortable, old fashioned, high ceilinged room
with heavy walnut furniture of the style of
forty years ago. An aged ingrain carpet was
on the floor, and a wreath of wax flowers
such as our grandmothers rejoiced in, hung,
set in a deep frame, on the wall. I thought
to myself that these were relics of departed
glories and of a day when there was money
to furnish the old hostel in the taste then in
vogue.*
—Effie Gladding, 1914

Lincoln Highway / old U.S. 50, Reese River valley and Shoshone Mountains, Lander County, Nevada.

One day in the parched lands of Nevada was much like another. Uphill, downhill; alkali dust, sand dunes. We climbed some fairly high passes and had more than the usual number of tire failures. Blowouts on two consecutive days could have been serious had not Al Jolson left a tire behind when he came to Eureka looking for a ghost-town setting for a picture. Al decided that the town wasn't quite dead enough to make a good ghost. But the tire was a perfect fit.
—Bellamy Partridge, 1913

Lincoln Highway / old U.S. 50 near Carroll Summit, Desatoya Mountains, Churchill County, Nevada.

All day thro rocky gorges around snaky precipitous trails such as you see in the movies where devil-may-care bandits ride, then across sage-brush mesas for hours at a stretch, and at the base of rocky mountain ranges where desperate rustlers hide in screenland. I was rather astonished that a Wells Fargo express coach didn't come rattling over the hills with the hero holding the heroine with one arm and shooting pursuing road agents out of their saddles with blank cartridges. It was great! Miles ahead we could see thin spirals of smoke hundreds of feet in the air. They seemed like camp fires or Indian signals, but later we caught up with the Fords that caused the dust clouds.
—James M. Flagg, 1925

Eastgate, Nevada.

Sand Springs pony express station,
Churchill County, Nevada.

Carson Lake bed, Churchill County, Nevada.

We came to a twenty-mile wide alkali as it was getting darker. A perfectly flat lake of alkali draw that looked like water at a distance, but when we reached it we only saw faint auto tracks. We just had to make Austin that evening as we had no food with us except some hot raisins in the trunk behind.
—James M. Flagg, 1925

Approaching the Sierra Nevada near Truckee, California.

Donner Pass, California; 1915 road foreground, 1925 road at left, Southern Pacific Railroad at top.

Many times I thought of the "Forty-niners," as we saw the sign, "Overland Trail." In coming along the Lincoln Highway, we are simply traversing the old overland road along which the prairie schooners of the pioneers passed. How much heart-ache, heart-break, and hope deferred this old trail has seen! I think of it as we bowl along so comfortably over the somewhat rough but yet very passable road.
—Effie Gladding, 1914

Donner Pass Bridge, completed 1925.

Abandoned Lincoln Highway near Dutch Flat, California.

Victory was in sight. We had passed the worst of our road problems, and the heat too. All around us were mountain peaks and each time we stopped the views became more extensive and more gorgeous. Trees were plentiful and greener than we had seen for a long time. It was incredible that so soon after leaving that arid, barren desert we could be refreshed in the cool shade of towering evergreens. That was balm to our bodies and spirits. Even the inarticulate Maxwell appeared to echo our sensations.

Majestic sugar pines, Douglas firs and redwoods lined our road on both sides. What a land! What mountains! What blue skies and clear, sparkling water! Our hearts leapt within us. None of us had ever seen the like—and we loved it. We almost chirped as we exclaimed over the grandeur that surrounded us on all sides. We started talking over plans when the trip was completed.
—Alice Ramsey, 1909

Sacramento Valley near Woodbridge, California.

Getting away early the next morning we ran down the warm Sacramento Valley. Fruit trees were heavily laden; we saw our first glimpses of apricots, pears, peaches, grapes, olives, oranges and apples. The boys hopped out and we had a nice armful of the latter. Dan and I blocked up the front spring, which had two broken blades; Alice and Margaret slept the while; Paschal scared up a deer, which made a clean leap across the road, one touch of the ground, over the fence and was seen no more.
—D. Norman Longaker, 1923

1915 Lincoln Highway near Altamont Pass, Diablo Range, California.

Summit Garage, Altamont, California.

ALTAMONT
N.Y.   S.F.
3284   47
Pop. 80. Alt. 737 feet. Alameda County.
One hotel. Local speed limit 25 miles per
hour, enforced. Route marked through town
and county. Extensive road improvement.
Two railroads, 1 general business place, 1
express company, 1 telegraph company, 1
public school, electric lights.
—Complete Official Road Guide of the
*   Lincoln Highway, 1916*

1928 Lincoln Highway marker at California Street and California Highway 1, San Francisco. The end of the Lincoln Highway was a mile west of here overlooking the Pacific Ocean.

As ill luck would have it, we had a blowout on one of San Francisco's busiest streets. A crowd collected quickly; and we were favored with various bits of advice concerning tires—all of which was typical of those days. During the course of this tire changing conference, one of the conferees spotted our Maryland license plates. "Boys," he said significantly, "you're a long ways from home!"

But later, as we stopped our faithful little cart at the Presidio, and gazed beyond the Golden Gate, out across the seemingly limitless expanse of the Pacific, we did not care if we were "a long ways from home." Our car had seen us through a thousand hazards in the crossing; and we knew that it would take us safely back, if we gave the command.
—William Starr, 1915
The Ohio Motorist

house. Late at night, when there is a soft breeze from the north, I can sometimes hear the whine of truck tires making time for Chicago or the West Coast.

Early one recent fall I made a hurried trip to San Francisco on that road. This was to be a true road trip: a direct two-and-a-half-day drive from home in Iowa City to San Francisco, twenty-four hours in the city, and a flight home. This adventure rose out of a young cousin's need for a co-driver on his long trip back to college in the Bay Area, my interest in seeing an old friend on Powell Street, and my hopes of finding a 1928 Lincoln Highway marker in the middle of San Francisco.

I had received a letter in weeks past from a man in Sacramento who had asked me if I had ever seen the Lincoln Highway marker in the west part of San Francisco. I had not. I called him eagerly for more details because my explorations of that city had yielded little in the way of Lincoln Highway significance, all signs of the highway's official end point at the Palace of the Legion of Honor having long since been removed. He told me that he had seen it within the last year or so and that he recalled it stood dark and overlooked by the curb at the intersection of Lake Street and Park Presidio Drive. While the trip west with my cousin would be pleasant, and the opportunity to see an old friend quite fun, my real reason for going was to find that marker.

My cousin and I left Iowa City early on a Friday morning, the station wagon piled high with the stuff of college, a motor scooter mounted on the back bumper. I carried a road atlas, a good street map of San Francisco, a couple changes of clothes, a notebook, and a bag of camera stuff. It was to be an unusual trip for me. While I do "take the I-road" quite often, it is seldom for more than a few hours. Beyond a certain point, no matter how urgent my arrival someplace may be, I inevitably veer off to some two-lane where I can gather my wits. For me, driving the interstate for long hours is too much like

playing the same song over and over again; the senses grow numb, and no matter how beautiful the passing countryside, my view turns to the speedometer and the traffic ahead. My cousin agreed. He had used his computer to generate a log from the notes of his earlier trips. It was broken down state by state and arranged by exit numbers, listing all the Standard stations, McDonald's, and Motel 6s. A log for rapid, undistinguished transit. It was called "I80: Iowa City to San Francisco via Boredom."

As with any road trip, the pace of travel was measured in hours at the wheel. At the legal limit, we ate up the miles in short order. A noon meal stop at Council Bluffs, Iowa, taught me to handle a Big Mac and drive an Oldsmobile at the same time; supper was at another McDonald's in Gothenburg, Nebraska. I could swear that the same young woman had waited on me a few hours earlier in Council Bluffs.

Some thirty miles on west near North Platte, our singing tires picked up the main trail of Oregon and California wagon pioneers. For the next sixty miles to Brule we stayed within a mile or so of this great route, but of the passage of many thousands of westbound emigrants, nothing could be seen in the evening light. As exit number 117, the off ramp for Brule flashed by, I recalled an earlier trip when I had climbed California Hill near there. I had sat for some time in the ruts left by thousands of wagons more than a century ago. A storm had been building to the west that day, and I had thought how the rain would have cooled the oxen after the long pull out of the South Platte valley.

At 10:30 we made Laramie and pulled into a Motel 6 just off the highway. In bed, I heard other late arrivals outside and thought of campfires back along the Platte River, campfires that had lit the prairie wilderness and had shimmered on the white canvas tops of overland wagons.

In the morning, McDonald's in Laramie was not yet open when we departed, so I suggested we drive to a little truck-stop cafe up the road at Walcott, where Interstate 80 and U.S. 30 rejoined. We burst through patches of ground fog in the low places as we drove west. The aspen on Elk Mountain were turning, and the radio talked of snow for the northern mountains.

The cafe at Walcott evoked a familiar rhythm from earlier trips, and though my cousin appeared eager to hurry along, I bought a paper and we both sat and read while awaiting eggs cooked by a person rather than a machine. As the waitress filled my coffee cup, she read a story over my shoulder—something or other about smoking—and issued a loud and certain complaint about people who light up but complain that they cannot afford other things. My cousin grew a bit uneasy at the outburst, but soon a good-natured cafe-wide conversation developed among ranchers, truck drivers, and the help. That's never happened at McDonald's.

The day passed at nearly a mile a minute; we played tapes, listened to the radio, and talked on the CB radio, a novelty to me which was sometimes helpful, sometimes interesting, and often obscene. I ticked off the places when I had stopped for the night on my previous trip west, when I had made my slow but deliberate way to San Francisco, photographing, looking for old routes, observing. We were doing in less than three days what had taken me more than three weeks in the spring.

I remembered camping at a favorite little campground near Potter, Nebraska, a place where Carol and I often stop: "Buffalo Bend Campground, Where the West Peters Out." I recalled standing on the bluff nearby at Point of Rocks and an icy May wind whistling through my jacket as I steadied a tripod. A few nights later I was in Medicine Bow at the old Virginian Hotel, made famous by western author Owen Wister.

I ate supper in the dining room that evening, called home on the pay phone in the lobby, and just as darkness settled in, I went upstairs and picked a room from among several standing with doors open: room number 5. How many automobile adventurers had stayed in this room before me? During the busy days on the Lincoln, this place would have been full every night during the summer; tonight I was the only guest. The room was warm, so I opened the window and stuck a boot under the heavy sash to hold it up. As I lay in bed, I could hear the kitchen crew downstairs banging pots and running water, finishing up the evening chores, then silence when they went home. Outside it was quiet, except for a diesel truck idling to itself across the road, its driver inside the bar attached to the hotel. Occasionally a car would pass, and from time to time a Union Pacific freight would crack the stillness, but when the truck eventually pulled away into the night, silence covered southern Wyoming like a blanket. And through the open window drifted the scent of the May night, the faint smell of sage, as intoxicating as anything the trucker had imbibed.

My cousin and I rolled through Salt Lake City with no stops for water, extra food, tow ropes, or new tires, and out onto the salt desert. West of Knolls I described some of the trials of travel years ago on the white salt—the Donner party, the miles between water, the endless trip by stage, Thomas's mudhole at Fish Springs—but the traffic was too heavy, the horizon of white was too distant, blocked by other vehicles. And my heart wasn't in it; I was beginning to feel like L. B. Miller. All that mattered was getting there. We crossed the salt desert and reached Wendover in forty-five minutes.

As we angled around the north ends of the mountain ranges in Nevada along the old Victory Highway route, I thought of my friend in San Francisco. We had known each other in Iowa City for several years, and on the night before he moved west, we had sat over a beer and talked of old times. At one point a puzzled and fearful look crossed his face and he said: "You know, I haven't the faintest idea about how to get to San Francisco. What roads do you take, what towns do you pass? I'll have to get a map."

"No," I said, "you won't need a map. Just get on Interstate 80 and drive until you see a big city and an ocean. Then, stop and ask what ocean it is. If they say Atlantic, turn around and go clear back. If they say Pacific, look for an apartment." We laughed, and he left the next day and found San Francisco.

My cousin and I stayed in Reno the second night out. The Motel 6 was full, so we found another place, one more interesting for being more seedy. In the morning we started early for the final miles to the coast. We climbed to the Sierra Nevada climax and got a brief glimpse of Donner Pass proper as we angled some distance to the north. No snow yet. Just beyond the summit we began to pick up coast city radio stations. In a few short hours we were in San Francisco, where we navigated the big wagon through San Francisco traffic. With the scooter on the back it looked like the Titanic with lifeboats.

I hurried through lunch with my friend, and since he had work to finish, I borrowed his car and drove to find the marker, only a mile from the ocean, one of the very last that Gael Hoag had located and marked in his field book.

But the marker was gone. A fresh panel of sidewalk occupied the spot where the post should have been. I was too late. I walked Lake Street in either direction for a block to make sure. Nothing. Convinced that the trip had been in vain, but nagged by a doubt that Lake Street ever was the Lincoln, I walked a block south and found the marker, serene and mossy, standing near the corner of Park Presidio Drive and California Street, the intersection of the coastal highway—California 1, the westernmost highway in the country—and the Lincoln Highway.

I hurried to photograph it before something happened to it, but laughed at myself. It had been here since 1928; it wouldn't be hit by a bus or car in the hour or two I was here. As I stood across the street watching, people came and went to board buses that stopped in front of it. Many people leaned on it or tried to rest packages on its pointed top, and one fellow tried to sit on it, but no one looked at it. While I stood and photographed, often waiting for traffic to clear in order to get an unobstructed view, a middle-aged woman bearing shopping bags saw my interest in this old concrete post and stopped to ask about it. She told me she walked by it several times every day and had done so for years but had never known what it stood for.

When I told her about it, she was skeptical. "You mean this is the end of a highway that went from coast to coast—from New York to here, right here on this street?" I assured her it did, that it passed at her feet and that it officially ended a mile west at Lincoln Park. I suggested she go across and read the medallion set into the post. Hoisting her grocery sacks, she crossed California Street and bent to read the lettering around the bust of Lincoln on the small bronze inset. Satisfied, she turned, smiled, and made a forearm wave so as not to drop her groceries, and walked toward home. The medallion on the old marker says, "This Highway Dedicated to Abraham Lincoln."

At my feet was the westernmost mile of the Lincoln Highway. To the east on this same path were the places that had crowded the memories of thousands of travelers. Miles to the east is the place at the west edge of Rochelle where Emily Post stood and waited for the soil of Illinois to dry enough to become a highway; the place where Bellamy Partridge saw the Great Plains for the first time; the spot where Bill's Place once stood in the mountains of Pennsylvania; and the exact place where the West began for every motorist. Along this highway is a place where Harry Ostermann's Packard slipped, rolled over, and cost him his life; somewhere is an old ranch in Nevada where Alice Ramsey breakfasted on lamb chops and chocolate cake; and someplace in Wyoming is a high, windswept plain where Henry Joy once stood with a cup in his hand and smelled bacon frying over a sagebrush campfire. He must have thought of the mud he and his companions had crossed, and he must have thought about paved roads for the nation.

Place-Names along the Lincoln Highway, 1915–1916

Many of these are towns and cities; quite a few are tiny villages. A handful are outposts along the railroad—section houses or telegraph stations. Others are simply ranches, and two or three are simply landmarks.

NEW YORK
New York City

NEW JERSEY
Jersey City
Newark
Elizabeth
Rahway
Iselin
Menlo Park
Metuchen
New Brunswick
Highland Park
Franklin Park
Kingston
Princeton
Lawrenceville
Trenton

PENNSYLVANIA
Oxford Valley
Glen Lake
Langhorne
La Trippe
Bustleton
North Philadelphia Station
Philadelphia
Overbrook
Ardmore
Bryn Mawr
Wayne
Berwyn
Paoli
Whitford
Downington
Thorndale Station

Coatsville
Sadsburyville
Mt. Vernon
Gap
Kinzers
Leaman Place
Paradise
Soudersburg
Lancaster
Mountville
Columbia
Wrightsville
York
Abbottstown
New Oxford
Gettysburg
Seven Stars
McKnightstown
Cashtown
Grafenburg
Caledonia Park
Fayetteville
West Fayetteville
Chambersburg
St. Thomas
Fort Loudon
McConnellsburg
Harrisonville
Breezewood
Everett
Mt. Dalls
The Willows
Bedford
Wolfsburg
Schellsburg
Buckstown

Kanter P.O.
Stoystown
Jenners
Jennerstown
McLaughlintown
Ligonier
Youngstown
Greensburg
Grapeville
Adamsburg
Irwin
Jacksonville
East McKeesport
Turtle Creek
East Pittsburgh
Wilkinsburg
Pittsburgh
Leetsdale
Fairoaks
Ambridge
Economy
Legionville
Logans
Baden
Conway
Freedom
Rochester
Bridgewater
Beaver
Esther
Ohioville
Smith's Ferry

OHIO
East Liverpool
Lisbon
Hanoverton

Kensington
East Rochester
Minerva
Robertsville
Osnaburg
Canton
Massillon
West Brookfield
East Greenville
Dalton
East Union
Wooster
Jefferson
New Pittsburg
Rowsburg
Ashland
Mansfield
Ontario
Galion
Bucyrus
Nevada
Upper Sandusky
Forest
Dunkirk
Dola
Ada
Lima
Gomer
Delphos
Van Wert

INDIANA
Fort Wayne
Churubusco
Merriam
Wolf Lake
Kimmell

Ligonier
Benton
Goshen
Elkhart
Osceola
Mishawaka
South Bend
New Carlisle
La Porte
Westville
Valparaiso
Deep River
Merrillville
Schererville
Dyer

ILLINOIS
Chicago Heights
Joliet
Plainfield
Aurora
Mooseheart
Batavia
Geneva
De Kalb
Malta
Creston
Rochelle
Ashton
Franklin Grove
Nachusa
Dixon
Sterling
Morrison
Fulton

IOWA
Clinton
Elvira
De Witt
Grand Mound
Calamus
Wheatland
Lowden
Clarence
Stanwood
Mechanicsville
Lisbon
Mt. Vernon
Marion
Cedar Rapids
Belle Plaine
Chelsea
Gladstone
Tama
Montour
Butlerville
Le Grand
Marshalltown
State Center
La Moille
Colo
Nevada
Ames
Ontario
Jordan
Boone
Ogden
Grand Junction
Jefferson
Scranton
Ralston
Glidden

Carroll
West Side
Vail
Denison
Arion
Dow City
Dunlap
Woodbine
Logan
Missouri Valley
Loveland
Honey Creek
Crescent
Council Bluffs

NEBRASKA
Omaha
Elkhorn
Waterloo
Valley
Fremont
Ames
North Bend
Rogers
Schuyler
Richland P.O.
Columbus
Duncan
Silver Creek
Clarks
Central City
Chapman
Grand Island
Alda
Wood River
Shelton
Gibbon

Kearney
Odessa
Elm Creek
Overton
Lexington
Cozad
Gothenburg
Brady
Maxwell
North Platte
Hershey
Sutherland
Paxton
Roscoe
Ogallala
Brule
Megeath
Big Springs
Chappell
Lodge Pole
Sunol
Sidney
Potter
Dix Station
Kimball
Bushnell

WYOMING
Pine Bluffs
Egbert
Burns
Archer
Cheyenne
Corlett Station
Borie Station
Otto Station
Granite Canyon Station

Buford
Sherman Hill
Laramie
Bosler
Cooper Lake
Lookout
Harper
Rock River
Medicine Bow
Allen Station
Hanna
Ft. Steele
Lakota
Granville
Rawlins
Creston Station
Latham Station
Wamsutter
Point of Rocks
Thayer Junction
Rock Springs
Green River
Bryan Station
Granger
Lyman
Ft. Bridger
Evanston

UTAH
Wyuta Station
Wasatch
Castle Rock
Emory Station
Main Forks
Coalville
Hoytsville
Wanship

Kimball's Ranch
Roach's Ranch
Salt Lake City
Pleasant Green
Ragtown
Garfield
Lakepoint
Milltown
Grantsville
Timpie Point
Iosepa
Brown's Ranch
Indian Ranch
Severe Farm
Orr's Ranch
County Well
Fish Springs
Callao
Ibapah

NEVADA
Tippett
Anderson's Ranch
Schellbourne
Magnuson's Ranch
McGill
East Ely
Ely
Lane
Copper Flat
Reipetown
Kimberly
Jake's Summit
Mooreman's Ranch
Rosevear's Ranch
White Pine Summit
Six Mile House

Pancake Summit	Hazen	Gold Run	Galt
Fourteen Mile House	Fernley	Colfax	Woodbridge
Pinto House	Wadsworth	Wymar	Stockton
Eureka	Derby	Applegate	French Camp
Austin	Vista	Auburn	Banta
New Pass Canyon	Sparks	New Castle	Tracy
Alpine Ranch	Reno	Penryn	Altamont
Eastgate	Verdi	Loomis	Livermore
Westgate		Rocklin	Hayward
Frenchman's Station	CALIFORNIA	Roseville	Oakland
Sand Springs	Truckee	Sacramento	San Francisco
Salt Wells	Donner	Elk Grove	
Grimes Ranch	Emigrant Gap	McConnell	
Fallon	Dutch Flat	Arno	

Notes on Sources

Any book of this sort relies a great deal on the piecing together of myriad bits of information from numerous sources in order to tell a complete story. Great piles of miscellaneous photocopied documents and correspondence, newspaper and magazine clippings, old road guides and maps, books about trips across the country by automobile, books about the automobile, and books about highways came to fill my office. Rather than burden the reader with the whole lot of it, here follow a few notes about my most important sources for those who wish to follow my trail.

As to the history of the highway and the association, *The Lincoln Highway,* published in 1935 by the Lincoln Highway Association, was a fundamental resource, but should be read as one would any authorized history. The papers of the Lincoln Highway Association provided a rich lode of original correspondence, bulletins to directors, maps, guides, and photographs. They are held at the University of Michigan Engineering and Transportation Library, Ann Arbor. Various newspaper articles, magazine stories, and other secondary sources fleshed out the account of the men who shepherded the Lincoln Highway through its early years.

The search for material about early auto touring began with a fine bibliography, *Autos across America: A Bibliography of Transcontinental Automobile Travel: 1903–1940,* Carey S. Bliss, Jenkins and Reese, 1982. Among the best and most useful travel accounts were:

By Motor to the Golden Gate, Emily Post, D. Appleton and Co., 1917
Truly Emily Post, Edwin Post, Funk and Wagnalls, 1961
Across the Continent by the Lincoln Highway, Effie Gladding, Brentano's, 1915
Fill'er Up, Bellamy Partridge, McGraw-Hill, 1952
Veil, Duster and Tire Iron, Alice Ramsey, Castle Press, 1961
The Family Flivvers to Frisco, Frederic F. Van de Water, D. Appleton and Co., 1927
The Good of It All, Thornton Round, Lakeside Printing Co., 1957
A Long Way from Boston, Beth O'Shea, McGraw-Hill, 1946
Boulevards All the Way—Maybe, James M. Flagg, George H. Doran Co., 1925

L. B. Miller's story is told in a rare pamphlet called *The Flight of the Gray Goose,* published by the Wills Sainte Claire Company. Useful for general background on auto touring were *Americans on the Road,* Warren Belasco, MIT Press, 1981, and *The Tourist,* John Jakle, University of Nebraska, 1985.

Much of the material about Henry Joy's trip west originates from a 1915 issue of *The Packard,* a company magazine. This and other material concerning Joy was provided by the Bentley Historical Library, University of Michigan. The story of Eisenhower's 1919 trip with the army is told in his *At Ease: Stories I Tell to Friends,* Doubleday and Co., 1967.

Several general books relating to the history of the automobile and auto travel in America were essential. *The Automobile in America,* Stephen W. Sears, American Heritage, 1977, is among the most readable. By John B. Rae are *The Road and the Car in American Life,* MIT Press, 1971, and *The American Automobile,* University of Chicago, 1965.

For locating the routes of the Lincoln and other named highways, I used old state highway maps from every year I could find, a reprint of the 1926 Rand McNally *Auto Road Atlas of the United States,* a scattering of touring guides including several *Official Automobile Blue Books,* Lincoln Highway Association guidebooks (including a reprint of the 1916 guide done by Lyn Protteau, P.O. Box 255185, Sacramento CA 95865), and many United States Geological Survey 1:250,000-scale maps. Of considerable value in tracing named highways was an unpublished typescript located at the American Automobile Association library in Falls Church, Virginia, called *Named Highways of the United States,* dated December 8, 1959.

I used the WPA guides of all twelve states along the Lincoln Highway to locate landmarks and to find earlier routes.

For material relating to the Lincoln Highway route prior to the coming of the automobile, I consulted several volumes of general U.S. transportation history, including *The History of Travel in America,* Seymour Dunbar, Tudor Publishing Co., 1937; *The Story of American Roads,* Val Hart, William Sloan Associates, 1950; and *Historic American Highways,* Albert C. Rose, American Association of State Highway Officials, 1953. In addition, I consulted many works that dealt with specific parts of the route. A partial list in order from east to west:

The First New York-Philadelphia Stage Road, James and Margaret Cawley, Associated University Presses, 1981

Indian Trails to Superhighways (Pennsylvania), William Shank, American Canal and Transportation Center, 1982

The California Trail, George R. Stewart, McGraw-Hill, 1962

The Story of the Pony Express, Glenn D. Bradley, Hesperian House, 1960

Donner Pass, George R. Stewart, Lane Book Co., 1964

Many miscellaneous manuscripts, publications, and periodicals from state historical societies in several states along the route also added to the story.

In the area of later highway culture, architecture, and construction, a handful of books stand out. For inspiration and a phylum system of roads, *U.S. 40,* George R. Stewart, Houghton Mifflin, 1953, was essential. *Highways in Our National Life,* Jean Labatut and Wheaton J. Lane, Princeton, 1950, was a useful source concerning engineering and construction matters. *The Verse by the Side of the Road,* Frank Rowsome, Stephen Greene Press, 1965, made entertaining reading on the Burma-Shave signs. *Main Street to Miracle Mile,* Chester H. Liebs, Little, Brown and Co., 1985, provided excellent background on changing roadside architecture.

Index